Dual Quaternions and Their Associated Clifford Algebras

Clifford algebra for dual quaternions has emerged recently as an alternative to standard matrix algebra as a computational framework for computer graphics. This book presents dual quaternions and their associated Clifford algebras in a new light, accessible to and geared toward the computer graphics community.

Collecting all the associated formulas and theorems in one place, this book provides an extensive and rigorous treatment of dual quaternions, as well as showing how two models of Clifford algebra emerge naturally from the theory of dual quaternions. Each section comes complete with a set of exercises to help readers sharpen and practice their understanding.

This book is accessible to anyone with a basic knowledge of quaternion algebra and is of particular use to forward-thinking members of the computer graphics community.

Dual Quaternions and Their Associated Clifford Algebras

Ronald Goldman

CRC Press
Taylor & Francis Group
Boca Raton London New York

CRC Press is an imprint of the
Taylor & Francis Group, an **informa** business

First edition published 2024
by CRC Press
2385 Executive Center Drive, Suite 320, Boca Raton FL 33431

and by CRC Press
4 Park Square, Milton Park, Abingdon, Oxon, OX14 4RN

CRC Press is an imprint of Taylor & Francis Group, LLC

ISBN: 9781032502960 (hbk)
ISBN: 9781032502977 (pbk)
ISBN: 9781003398141 (ebk)

DOI: 10.1201/9781003398141

Typeset in Palatino
by codeMantra

To my most immediate family: Jackie, April, Cody, Lloyd, and Pinky.

He proves by algebra that Shakespeare's ghost is Hamlet's grandfather.

James Joyce, *Ulysses*

Contents

Part II Clifford Algebras for Dual Quaternions

Preface

Dual quaternions are much more powerful, yet much more obscure than classical quaternions. Quaternions were introduced by Hamilton [14] to model rotations in 3-dimensions; dual quaternions were introduced by Clifford [1] to model rigid motions – translations and rotations —in 3 dimensions. Quaternion multiplication can be used to compute rotations, reflections, and perspective projections in 3-dimensions; dual quaternion multiplication can be used to compute translations as well as rotations and reflections along with perspective projections in 3-dimensions. Quaternions are vectors in 4-dimensions; dual quaternions are vectors in 8-dimensions. Quaternions can be represented by a pair of complex numbers; dual quaternions can be represented by a pair of quaternions.

Thus dual quaternions are both more powerful and more complicated than quaternions. These two factors are motivation enough for an extended rigorous study of dual quaternions. Moreover, recently several authors [5,11,12,13,25] have suggested that a Clifford algebra for dual quaternions is a more suitable framework for computer graphics than standard matrix algebra [8,15]. These claims too motivate a renewed interest in dual quaternions and their associated Clifford algebras.

This book is a sequel to my monograph *Rethinking Quaternions* [9], which also includes a discussion of a Clifford algebra for quaternions. Though this monograph on quaternions is not required reading for this sequel on dual quaternions, a thorough knowledge of quaternions is a prerequisite for understanding this book on dual quaternions.

Part I of this monograph is devoted to a thorough investigation of dual quaternions. Dual quaternions have already been written about by many authors, and I have borrowed extensively from the previous literature both classical and contemporary on this topic [1,3,4,10,18,19,20,23,24]. My goal here, however, is somewhat different: I want to present dual quaternions in a new light, accessible to and geared toward the computer graphics community. Therefore, in addition to presenting the algebra surrounding dual quaternions, I shall also show how to build on the intuitive geometric understanding of quaternions presented in [9] to develop a better intuitive understanding of dual quaternions. Moreover, in addition to deriving the formulas for rigid motions, I will also show here for the first time how to compute perspective and pseudo-perspective using dual quaternions.

Part II of this monograph is devoted to the study of two Clifford algebras associated to dual quaternions: the *plane model* and the *point model*. Although Part II begins with a brief review of the bare essentials of Clifford algebra, this initial chapter is not intended as a first introduction to Clifford algebra. Readers not at all familiar with Clifford algebra will be able to understand the subsequent material on Clifford algebras for dual quaternions, but for context readers would benefit immensely from some prior knowledge of Clifford algebra. For a good introduction to Clifford algebra, see for example [6,12,13].

The two Clifford algebras associated to dual quaternions have also been written about before [5,11,12,13,16,21,22,25], but some of the results are scattered throughout the literature; moreover, side-by-side comparisons are not readily available. My goals here are twofold: to collect in one place for easy learning and reference these two Clifford algebras, and to provide comparisons of the advantages and disadvantages of the strengths and weaknesses of each of these approaches to Clifford algebra for the dual quaternions. Rather than rehashing methods already in the literature, I will present a novel approach to these Clifford algebras. Once we have a thorough understanding of dual quaternions, I shall show how to derive easily, almost mechanically, geometric representations and algebraic formulas in these two Clifford algebras directly from representations and formulas of the corresponding geometry and algebra for the dual quaternions. This approach contrasts sharply with many standard presentations of these Clifford algebras, where the representations of geometry, though valid, are not well motivated, and the formulas in these algebras are derived *ab initio* without any direct reference to dual quaternions.

This book is divided into two parts: Part I deals exclusively with dual quaternions; Part II is devoted to the study of two Clifford algebras associated to dual quaternions. Readers interested only in dual quaternions can read Part I and avoid altogether the material on Clifford algebras presented in Part II. Exercises are provided after each section to help the reader master the mathematical techniques as well as to further illuminate the theorems by presenting alternative proofs.

Part I is organized in the following fashion. In order to place the dual quaternions in the context of the larger framework of mathematical algebras, Section 1 begins with a brief overview of some standard algebras and their corresponding dual algebras: real numbers and dual real number, complex numbers and dual complex numbers, quaternions and dual quaternions, octonions and dual octonions. Although the dual quaternions and octonions are both 8-dimensional algebras, the dual quaternions should not be confused with the octonions: the octonions are a nonassociative division algebra, whereas the dual quaternions are an associative algebra with zero divisors.

Section 2 commences the formal study of dual quaternions with the algebra underlying dual quaternions. Since dual quaternions are represented by pairs of quaternions, we begin with a brief review of quaternion algebra, just enough to understand the extension of quaternion algebra to dual quaternions. Then we introduce the basic formulas that form the foundation for the algebra of dual quaternions: the three conjugates and the dot product.

Geometry is next. Section 3 is devoted to representations for points, vectors, planes, and lines. Incidence relations, distance formulas, and intersection formulas are covered as well: some in the main body of the text, others in the exercises. The natural representations for lines in the space of dual quaternions are Plucker coordinates and dual Plucker coordinates, so for the uninitiated we review here as well these two representations for lines. Finally in Section 3, we discuss duality in the space of dual quaternions, a crucial topic for understanding our approach to deriving the formulas for perspective and pseudo-perspective in Section 7.

One of the primary goals of dual quaternions is to provide neat, alternative representations for some of basic transformations of computer graphics: translations, rotations, reflections, and perspective projections. Section 4 deals with how to apply unit dual quaternions to compute rigid motions: translations in arbitrary directions, rotations about arbitrary lines, and reflections in arbitrary planes. Along the way we prove Chasles' Theorem that every orientation preserving rigid motion is equivalent to a screw transformation, a single rotation about a fixed axis followed by a single translation in the direction of the axis of rotation.

Unit quaternions represent rotations in 4-dimensions. In Section 5, we show that unit dual quaternions represent rotations in 8-dimensions. We also show how to renormalize dual quaternions into unit dual quaternions to avoid distortions that could occur during rigid motions due to accumulated floating point errors.

One of the advantages of the quaternion representation for rotations over the matrix representation for rotations is the ease of interpolating between two rotations represented by unit quaternions using Spherical Linear Interpolation (SLERP). Similarly, Screw Linear Interpolation (ScLERP) can be applied to interpolate between two rigid motions represented by unit dual quaternions. Section 6 is devoted to deriving the formulas for Screw Linear Interpolation.

Perspective and pseudo-perspective are crucial in computer graphics for realistic rendering. In Section 7 duality is employed together with a dual quaternion representing translation to compute perspective and pseudo-perspective. Much of this material is new and is presented here for the first time.

Quaternions can be visualized by their projections into two mutually orthogonal planes in 4-dimensions. These projections can help us to visualize the results of multiplying arbitrary quaternions by unit quaternions. These visualizations in turn help to explain intuitively how quaternions can be used to compute rotations on vectors in 3-dimensions [8,9]. Similarly, dual quaternions can be visualized by their projections into four mutually orthogonal planes in 8-dimensions. In Section 8 we make dual quaternions visible by visualizing dual quaternions as projections into four mutually orthogonal planes in 8-dimensions. We then go on to explain intuitively why unit dual quaternions can be used to compute rigid motions in 3-dimensions. In particular we see that translations in 3-dimensions are represented by shears in 8-dimensions.

Matrices are typically used to represent transformations in computer graphics [8,15]. In Section 9 we compare and contrast the advantages and disadvantages of representations and computations with matrices to representations and computations with dual quaternions. We show that each method has some relative advantages and disadvantages. We also show how to convert between these two different representations for transformations.

In Section 10 we compare some properties of quaternions and dual quaternions and present as well some of the major insights that guided our intuition for the development of the theory of dual quaternions.

We close Part I in Section 11 with a summary for easy reference of our main formulas for algebra, geometry, duality, transformations, interpolation, and conversion between dual quaternions and matrices. Since we make occasional use in Part I of cross products, we also include an Appendix with some standard identities for cross products. These identities are derived in the Exercises of Part II, Chapter 1, Section 4.1, where we show how the cross product is related to the wedge product of Clifford algebra.

Part II contains three Chapters. Chapter 1 is a concise introduction to Clifford algebra, presenting only material needed to understand the Clifford algebras associated to dual quaternions. Chapter 2 presents what I call *the plane model of Clifford algebra*, and Chapter 3 presents what I call *the point model of Clifford algebra*. Chapter 3 ends with a side-by-side comparison of the strengths and weakness, the advantages and disadvantages of these two approaches to Clifford algebra for dual quaternions.

Chapter 1 is composed of four sections. Section 1 reviews the main goals of Clifford algebra, and Section 2 introduces the basic algebraic formalisms of Clifford algebra. Section 3 recalls the three main products in Clifford algebra: the Clifford product, the inner (dot) product, and the outer (wedge) product, and Section 4 presents duality (Hodge star), a major tool in our investigation of the Clifford algebras for dual quaternions.

Chapter 2 is devoted to the plane model of Clifford algebra. The basic algebraic formulas are presented in Section 1; geometry is described in Section 2. This model of Clifford algebra is called *the plane model* because planes are the basic geometric objects in this algebra (Section 2.1): lines are represented as the intersection of two planes (Section 2.3), points and vectors are represented as the intersection of three planes (Section 2.2). Rather than introduce the representations of planes, lines, points and vectors by fiat, we show how these representations arise naturally from the corresponding representations of planes, lines, points, and vectors in the space of dual quaternions. Incidence relations, distance formulas, and intersection formulas are covered as well in the same fashion, each derived from the corresponding formulas in the space of dual quaternions. Some of these formulas appear in the main body of the text, others in the exercises. Lines are naturally represented in this model as the intersection of two planes. But two points also determine a line. We introduce duality here (Section 2.4) to show how to represent lines as the join of two points.

One of the main powers of Clifford algebra is the computation of transformations represented by rotors and versors. In the plane model, rotors correspond to translations and rotations, versors correspond to reflections. Rather than start from first principles to derive the formulas for these transformations, we show in Section 3 how these formulas arise naturally and can be derived easily from the corresponding formulas in the space of dual quaternions. Perspective and pseudo-perspective using a translation rotor together with duality are presented in Section 3.4, and are again derived almost effortlessly from the corresponding formulas in the space of dual quaternions. Alternative derivations of all these transformations that do not rely on knowledge of the corresponding formulas for dual quaternions are provided in the exercises.

In Section 4 we list five of the major insights that guided our intuition for the study of the plane model of Clifford algebra, and in Section 5 we provide a summary for easy reference of our main formulas for the algebra, geometry, and transformations in the plane model of Clifford algebra.

We close Chapter 2 in Section 6 with a comparison of the corresponding formulas in the space of dual quaternions and in the plane model of Clifford algebra, and we compare as well the advantages and disadvantages of each of these two algebras.

Chapter 3 is devoted to the point model of Clifford algebra. Here we follow the same outline as in Chapter 2. The basic algebraic formulas are presented in Section 1; geometry is described in Section 2. This model of Clifford algebra is called *the point model* because points are the basic geometric objects in this algebra (Section 2.1): lines are represented as the join of two points (Section 2.4), planes are represented either as the join of three points or as the join of a point and two vectors (Section 2.2). Once again rather than introduce the definitions of points, vectors, planes, and lines by fiat, we show how these definitions arise naturally from the corresponding definitions of points, vectors, planes, and lines in the space of dual quaternions. Incidence relations, distance formulas, and intersection formulas are covered as well in the same fashion, each derived from the corresponding formulas in the space of dual quaternions. Again, some of these formulas appear in the main body of the text, others in the exercises.

Lines are naturally represented in this model as the join of two points. We introduce duality here (Section 2.3) to show how to represent lines in this model as the intersection of two planes.

Just as in the plane model, in the point model of Clifford algebra rotors correspond to translations and rotations, versors correspond reflections. Once again, we show in Section 3 how these formulas arise naturally and can be derived easily from the corresponding formulas in the space of dual quaternions. Perspective and pseudo-perspective using a translation rotor together with duality are presented in Section 3.4, and are derived almost effortlessly from the corresponding formulas in the space of dual quaternions. Alternative derivations of all these transformations that do not rely on knowledge of the corresponding formulas for dual quaternions are once again provided in the exercises.

In Section 4 we list seven of the major insights that guided our intuition for the study of the point model of Clifford algebra, and in Section 5 we provide a summary for easy reference of our main formulas for the algebra, geometry, and transformations in the plane model of Clifford algebra.

We close Chapter 3 in Section 6 with a comparison of the corresponding formulas in the point model and in the plane model of Clifford algebra, and we compare as well the advantages and disadvantages of each of these two Clifford algebras.

Why are there two models of Clifford algebra for dual quaternions? Neither model is completely satisfying. The plane model has a natural algebra, but an exotic geometry. Planes are the basic geometric elements in this model rather than points and vectors: points and vectors are generated from the intersections of three planes. In contrast, the point model has a natural geometry: points and vectors are the basic geometric elements, and planes are generated from the joins of points and vectors. But the algebra of the point model is exotic; an infinitesimal appears in the denominator of the square of one of the basis elements. Thus infinitesimals (or equivalently limits) often appear in computations involving the point model and must be discarded at the end of these computations. I leave it to the readers to decide which of these two Clifford algebras they prefer.

I did not come upon the ideas in this book in isolation; I learned from many other authors who preceded me. I initially learned about dual quaternions by reading the paper by Kavan et al. [19]. This paper first got me excited about the potential power of dual quaternions and eventually led me to write Part I of this monograph on dual quaternions. The material on ScLERP in Part I, Section 6 is largely an amplification of material in [18]. I learned about the plane model largely from [5,11,12,13,25], and I learned about the point model from [16,21,22]. I learned about the role and importance of duality in Clifford algebra from reading a preliminary version of [5]. I could not have written the sections on perspective and pseudo-perspective without an understanding of duality. Duality also plays a central role in the plane model for expressing a line as the join of two points, and in the point model for expressing a line as the intersection (meet) of two planes. I have borrowed many topics shamelessly from several other authors both past and present, most of whom are listed in the bibliography. My apologies to anyone I have inadvertently omitted. Of course, any errors in the text are solely my own. Signs are a particular tripping point in Clifford algebra; I trust I have finally gotten all of these many signs adjusted correctly.

To conclude: I would like to thank family and friends, colleagues and collaborators, students and teachers, novices and apprentices for their ubiquitous help and inspiration.

בָּרוּךְ אַתָּה יהוה, אֱלֹהֵינוּ מֶלֶךְ הָעוֹלָם, שֶׁהֶחֱיָינוּ וְקִיְּמָנוּ וְהִגִּיעָנוּ לַזְּמָן הַזֶּה.

July 18, 2022
Truckee, California

Author

Ronald Goldman is a Professor of Computer Science at Rice University in Houston, Texas. Professor Goldman received his B.S. in Mathematics from the Massachusetts Institute of Technology in 1968 and his M.A. and Ph.D. in Mathematics from Johns Hopkins University in 1973.

Professor Goldman's current research concentrates on the mathematical representation, manipulation, and analysis of shape using computers. His work includes research in computer-aided geometric design, solid modeling, computer graphics, subdivision, polynomials, and splines. His most recent focus is on the uses of quaternions, dual quaternions, and Clifford algebras in computer graphics.

Dr. Goldman has published over two hundred research articles in journals, books, and conference proceedings. He has also authored two books on computer graphics and geometric modeling: *Pyramid Algorithms: A Dynamic Programming Approach to Curves and Surfaces for Geometric Modeling* and *An Integrated Introduction to Computer Graphics and Geometric Modeling.* In addition, he has written an extended monograph on quaternions: *Rethinking Quaternions: Theory and Computation.*

Dr. Goldman is currently an Associate Editor of *Computer Aided Geometric Design.*

Before returning to academia, Dr. Goldman worked for ten years in industry solving problems in computer graphics, geometric modeling, and computer-aided design. He served as a Mathematician at Manufacturing Data Systems Inc., where he helped to implement one of the first industrial solid modeling systems. Later he worked as a Senior Design Engineer at Ford Motor Company, enhancing the capabilities of their corporate graphics and computer-aided design software. From Ford he moved on to Control Data Corporation, where he was a Principal Consultant for the development group devoted to computer-aided design and manufacture. His responsibilities included database design, algorithms, education, acquisitions, and research.

Dr. Goldman left Control Data Corporation in 1987 to become an Associate Professor of Computer Science at the University of Waterloo in Ontario, Canada. He joined the faculty at Rice University in Houston, Texas as a Professor of Computer Science in July 1990.

Part I

Dual Quaternions

1.1

Algebras and Dual Algebras

Dual quaternions are a dual algebra. An *algebra* is a vector space together with a rule for multiplication. A *dual algebra* is similar to an algebra, but the coefficients of a dual algebra are not real numbers; rather the coefficients of a dual algebra are dual real numbers (see below).

To introduce the dual quaternions, we shall briefly review here for context and contrast four related algebras and their associated dual algebras: real numbers, complex numbers, quaternions, and octonions.

1. *Real Numbers*
   In 1-dimension we have the real numbers. Geometrically each real number is associated with a unique point on a line. Multiplication is associative, commutative, and distributes through addition.
2. *Complex Numbers*
   In 2-dimensions we have the complex numbers. Geometrically each complex number is associated with a unique point in the plane—that is, with a unique pair of real numbers:

$$(x,y) \leftrightarrow z = x + yi \quad i^2 = -1.$$

   Algebraically the symbol i commutes with every real number and multiplication distributes through addition. These rules lead to the following formula for the product of two complex numbers:

$$(a + bi)(x + yi) = (ax - by) + (ay + bx)i.$$

   Multiplication for complex numbers is both associative and commutative.
3. *Quaternions*
   In 4-dimensions we have the quaternions. Geometrically each quaternion is associated with a unique vector in 4-dimensions— that is, with a unique pair of complex numbers or equivalently with four real numbers:

$$(z_1, z_2) \leftrightarrow q = z_1 + z_2 j = (x_1 + y_1 i) + (x_2 + y_2 i)j \quad j^2 = -1.$$

DOI: 10.1201/9781003398141-2

3

Algebraically the symbol j commutes with every real number and anticommutes with the complex number i. To designate the product ij, we introduce a new parameter k to represent a new direction in space and set $k = ij$. Notice that

$$k^2 = (ij)(ij) = -i(j^2)i = i^2 = -1.$$

Thus any quaternion q can be written as

$$q = x_1 + y_1 i + x_2 j + y_2 k,$$

where

$$i^2 = j^2 = k^2 = -1.$$

Again multiplication distributes through addition, so using the identities $zj = jz^*$ and $jz = z^* j$, where $z^* = x - yi$ denotes the complex conjugate of $z = x + yi$, we have the following formula for the product of two quaternions (see Exercise 5):

$$(w_1 + w_2 j)(z_1 + z_2 j) = (w_1 z_1 - w_2 z_2^*) + (w_1 z_2 + w_2 z_1^*) j.$$

Alternatively, adopting the notation of [9], which we shall use throughout this text, we can express a quaternion as a sum $mO + v$, where m is a real number, O is the point at the origin in 3-dimensions and $v = v_1 i + v_2 j + v_3 k$ is vector in 3-dimensions. Now we have the following classical formula for the product of two quaternions [9]:

$$(mO + u)(nO + v) = (mn - u \cdot v)O + mv + nu + u \times v.$$

Notice that O is the identity for quaternion multiplication. Multiplication for quaternions is associative but not commutative.

4. *Octonions*

In 8-dimensions we have the octonions. Geometrically each octonion is associated with a unique vector in 8-dimensions—that is, with a unique pair of quaternions or equivalently with eight real numbers:

$$(q_1, q_2) \leftrightarrow q_1 + q_2 l = (a_1 + b_1 i + c_1 j + d_1 k) + (a_2 + b_2 i + c_2 j + d_2 k)l \quad l^2 = -1.$$

The new symbol l commutes with every real number and anticommutes with the quaternions i, j, k. Multiplication for octonions, like multiplication for quaternions, is not commutative, but multiplication does distribute through addition. One must be careful, however, because multiplication for octonions is not associative, so using the identities $ql = lq^*$ and $lq = q^*l$, where $q^* = a - bi - cj - dk$ denotes the quaternion conjugate of $q = a + bi + cj + dk$, is not enough to define multiplication for octonions (see Exercise 6). In fact, we have the following formula for the product of two octonions [2]:

$$\left(p_1 + p_2 l\right)\left(q_1 + q_2 l\right) = \left(p_1 q_1 - q_2^* p_2\right) + \left(q_2 p_1 + p_2 q_1^*\right) l.$$

Notice the similarity of octonion multiplication to quaternion multiplication. Nevertheless, multiplication for octonions is not associative (see Exercise 6).

The real numbers, complex numbers, quaternions, and octonions in dimensions $1, 2, 4, 8$ are *division algebras*: every nonzero element has a multiplicative inverse. This progression, however, stops here: no 16-dimensional division algebra can be constructed in this way, or in any other way.

Dual quaternions, like octonions, can be represented by a pair of quaternions. Thus, the dual quaternions, like the octonions, are an algebra for an 8-dimensional vector space. Nevertheless, the dual quaternions are not the same as the octonions. Octonions are hard to work with because octonion multiplication is not associative. We shall not have any further use for octonions in this text. For readers who want to learn more about octonions—their idiosyncratic algebra and their potential geometric applications—see [2]. In contrast to the multiplication for octonions, we shall see shortly that multiplication for dual quaternions is associative, although like quaternion multiplication the product rule for dual quaternions is not commutative.

The dual quaternions can be constructed in a way similar to the octonions, but with one important difference: the rule $l^2 = -1$ for the octonions is replaced by the rule $\varepsilon^2 = 0$ for dual quaternions, where $\varepsilon \neq 0$ is a new parameter.

Just like the rule $i^2 = -1$ leads to a progression from real numbers to complex numbers to quaternions to octonions, the rule $\varepsilon^2 = 0$ leads to a progression from dual real numbers to dual complex numbers to dual quaternions to dual octonions. Here is how this progression works.

5. *Dual Real Numbers*

In 2-dimensions we have the dual real numbers. Geometrically each dual real number, like each complex number, is associated with a unique point in the plane—that is, with pair of real numbers:

$$\left(x, y\right) \leftrightarrow d = x + y\varepsilon \qquad \varepsilon^2 = 0.$$

Algebraically, the new symbol $\varepsilon \neq 0$ commutes with every real number and multiplication distributes through addition. But, in contrast to the complex numbers, we have the following formula for the product of two dual real numbers:

$$(a + b\varepsilon)(x + y\varepsilon) = ax + (ay + bx)\varepsilon.$$

Notice how this product formula mimics the addition formula for fractions:

$$b/a + y/x = (ay + bx)/ax.$$

Thus the dual real numbers are a mechanism for converting addition of fractions into multiplication of ordered pairs. Multiplication for dual real numbers, like addition for fractions, is both associative and commutative. Like complex numbers, the dual real number $x + y\varepsilon$ represents the point in the plane with coordinates (x, y), but multiplication by ε no longer represents rotation by $90°$ (see Exercise 1). We shall shortly see that these dual real numbers serve as the coefficients for the dual quaternions.

6. *Dual Complex Numbers*

In 4-dimensions we have the dual complex numbers. Geometrically each dual complex number is associated with a unique vector in 4-dimensions—that is, with a pair of complex numbers or equivalently with four real numbers:

$$(z_1, z_2) \leftrightarrow z_1 + z_2\varepsilon = (x_1 + y_1 i) + (x_2 + y_2 i)\varepsilon \quad \varepsilon^2 = 0.$$

Algebraically the new symbol $\varepsilon \neq 0$ commutes with every complex number and multiplication distributes through addition, so we have the following formula for the product of two dual complex numbers:

$$(w_1 + w_2\varepsilon)(z_1 + z_2\varepsilon) = w_1 z_1 + (w_1 z_2 + w_2 z_1)\varepsilon.$$

Notice again how this product formula mimics the addition formula for complex fractions:

$$w_2 / w_1 + z_2 / z_1 = (w_1 z_2 + w_2 z_1) / w_1 z_1.$$

Multiplication for dual complex numbers, like addition for complex fractions, is both associative and commutative, but we shall not make any use of these dual complex numbers here.

7. *Dual Quaternions*

In 8-dimensions we have the dual quaternions. Geometrically each dual quaternion is associated with a unique vector in 8-dimensions—that is, with a pair of quaternions or equivalently with eight real numbers or four dual real numbers:

$$\left(q_1, q_2\right) \leftrightarrow q_1 + q_2\varepsilon = (a_1 + b_1 i + c_1 j + d_1 k)$$

$$+ (a_2 + b_2 i + c_2 j + d_2 k)\varepsilon$$

$$= (a_1 + a_2\varepsilon) + (b_1 + b_2\varepsilon)i + (c_1 + c_2\varepsilon)j$$

$$+ (d_1 + d_2\varepsilon)k \quad \varepsilon^2 = 0.$$

Algebraically the new symbol $\varepsilon \neq 0$ commutes with every quaternion and multiplication distributes through addition, so we have the following formula for the product of two dual quaternions:

$$(p_1 + p_2\varepsilon)(q_1 + q_2\varepsilon) = p_1 q_1 + \left(p_1 q_2 + p_2 q_1\right)\varepsilon, \tag{1.1}$$

where the multiplication on the right-hand side is quaternion multiplication. This product formula seems to mimic a formula for addition for quaternion fractions, but since quaternion multiplication is not commutative, the notions of denominators and common denominators for quaternions are more subtle. Nevertheless, we have the following formula:

$$p_2 p_1^{-1} + q_1^{-1} q_2 = p_2\left(p_1^{-1} q_1^{-1}\right)q_1 + p_1 (p_1^{-1} q_1^{-1})q_2.$$

Unlike octonion multiplication, multiplication for the dual quaternions is associative, but like quaternion multiplication, multiplication for the dual quaternions is not commutative. Notice that the coefficients for the dual quaternions relative to the quaternions are the dual real numbers. Indeed let $d_1 = a_1 + b_1\varepsilon$ and $d_2 = a_2 + b_2\varepsilon$ be dual real numbers and let $q = q_1 + q_2\varepsilon$ be a dual quaternion. Then it is straightforward to verify that

$$(d_1 d_2)q = d_1 (d_2 q) \quad \text{(see Exercise 7)}.$$

Thus, we shall see in Section 3 that, unlike quaternions, for dual quaternions dot products and norms—angles and lengths—are not real numbers but rather are dual real numbers. Moreover, although the quaternions are contained as a subalgebra of the dual quaternions—the subalgebra where the coefficient of ε is zero—this subalgebra is relative to the real numbers, not relative to the dual real numbers, since the product of a dual real number and a quaternion is not another quaternion.

The only associative division algebras for real vector spaces are in dimensions 1, 2, and 4 corresponding to the real numbers, the complex numbers and the quaternions [17]. The octonions do not violate this theorem because octonion multiplication is not associative. The dual quaternions do not violate this theorem because the dual quaternions are not a division algebra: $\varepsilon^2 = 0$, so $\varepsilon \neq 0$ has no inverse.

Exercises

1. Show that geometrically in the space of dual real numbers multiplication by ε represents the composition of projection onto the x-axis and rotation by $90°$.

2. Show that for complex numbers:
   a. $e^{ti} = \cos(t) + i\sin(t)$ (Hint: Use the Taylor expansion for e^x.)
   b. $e^{ti}(x + yi) = x\cos(t) - y\sin(t) + (y\cos(t) + x\sin(t))i$
   c. Conclude from part b that the map $(x + yi) \mapsto e^{ti}(x + yi)$ represents a rotation in the xy-plane.

3. Show that for the dual real numbers:
   a. $e^{t\varepsilon} = 1 + t\varepsilon$ (Hint: Use the Taylor expansion for e^x.)
   b. $e^{t\varepsilon}(x + y\varepsilon) = x + (y + tx)\varepsilon$
   c. Conclude from part b that the map $(x + y\varepsilon) \mapsto e^{t\varepsilon}(x + y\varepsilon)$ represents a shear in the xy-plane. (Compare to Exercise 2.)

4. Show that:
   a. Multiplication for the dual real numbers is associative and commutative.
   b. Multiplication for the dual complex numbers is associative and commutative.
   c. Multiplication for the dual quaternions is associative but not commutative.

5. Let $z^* = x - yi$ denote the complex conjugate of $z = x + yi$. Using the formula for quaternion multiplication, show that:
   a. $zj = jz^*$ and $jz = z^*j$
   b. $(w_1 + w_2 j)(z_1 + z_2 j) = (w_1 z_1 - w_2 z_2^*) + (w_1 z_2 + w_2 z_1^*)j$.

6. Let $q^* = w - xi - yj - zk$ denote the quaternion conjugate of $q = w + xi + yj + zk$. Using the formula for octonion multiplication, show that:

   a. $ql = lq^*$ and $lq = q^*l$

   b. Using part *a*, show that if octonion multiplication were associative, then we would have

   $$\left(p_1 + p_2 l\right)\left(q_1 + q_2 l\right) = \left(p_1 q_1 - p_2 q_2^*\right) + \left(p_1 q_2 + p_2 q_1^*\right)l.$$

   c. Conclude from part *b* that octonion multiplication is not associative.

7. Let $d_1 = a_1 + b_1 \varepsilon$ and $d_2 = a_2 + b_2 \varepsilon$ be dual real numbers and let $q = q_1 + q_2 \varepsilon$ be a dual quaternion. Show that $(d_1 d_2)q = d_1(d_2 q)$.

8. Let $q = q_1 + q_2 \varepsilon$ be a dual quaternion with $q_1 \neq 0$. Show that $q^{-1} = q_1^{-1}\left(1 - q_2 q_1^{-1}\varepsilon\right)$.

9. (*Hyperbolic Numbers*) Each *hyperbolic number* is associated with a unique point in the plane—that is, with a pair of real numbers:

   $$\left(x,y\right) \leftrightarrow w = x + yh \quad h^2 = 1.$$

   The new symbol h commutes with every real number and multiplication is associative and distributes through addition. Show that:

   a. $\left(a + bh\right)\left(x + yh\right) = \left(ax + by\right) + \left(ay + bx\right)h.$

   b. Multiplication for the hyperbolic numbers is commutative.

   c. The map $\left(x + yh\right) \mapsto h\left(x + yh\right)$ reflects points in the line $y = x$.

   d. $e^{h\theta} = \cosh\left(\theta\right) + h\sinh\left(\theta\right).$

   e. $e^{h\theta}\left(x + yh\right) = \left(x\cosh\left(\theta\right) + y\sinh\left(\theta\right)\right) + \left(x\sinh\left(\theta\right) + y\cosh\left(\theta\right)\right)h.$

   f. Conclude from part *e* that the map $\left(x + yh\right) \mapsto e^{h\theta}\left(x + yh\right)$ represents a scissor shear in the xy-plane. (Compare to Exercises 2 and 3.)

10. Consider the polynomial $t^2 - \gamma$. Show that:

    a. $t^2 - \varepsilon$ has no roots in the dual complex numbers.

    b. $t^2 - i$ has 2 roots in the complex numbers.

    c. $t^2 - h$ has 4 roots in the hyperbolic complex numbers.

11. Show that:

    a. The complex numbers are isomorphic to the 2×2 matrices

    $$\left\{ \begin{pmatrix} a & b \\ -b & a \end{pmatrix} \mid a,b \in \mathbb{R} \right\}.$$

    In particular, show that $1 \leftrightarrow \begin{pmatrix} 1 & 0 \\ 0 & 1 \end{pmatrix}$ and $i \leftrightarrow \begin{pmatrix} 0 & 1 \\ -1 & 0 \end{pmatrix}$.

b. The dual real numbers are isomorphic to the 2×2 matrices

$$\left\{ \begin{pmatrix} a & b \\ 0 & a \end{pmatrix} \mid a,b \in \mathbb{R} \right\}.$$

In particular, show that $1 \leftrightarrow \begin{pmatrix} 1 & 0 \\ 0 & 1 \end{pmatrix}$ and $\varepsilon \leftrightarrow \begin{pmatrix} 0 & 1 \\ 0 & 0 \end{pmatrix}$.

c. The hyperbolic numbers are isomorphic to the 2×2 matrices

$$\left\{ \begin{pmatrix} a & b \\ b & a \end{pmatrix} \mid a,b \in \mathbb{R} \right\}.$$

In particular, show that $1 \leftrightarrow \begin{pmatrix} 1 & 0 \\ 0 & 1 \end{pmatrix}$ and $h \leftrightarrow \begin{pmatrix} 0 & 1 \\ 1 & 0 \end{pmatrix}$.

12. Continuing with Exercise 11, let $\lambda = i$, ε, h, and let $p = a + b\lambda$. Define

$$p^* = a - b\lambda = \text{the conjugate of } p$$

$$|p|^2 = pp^* = (a+b\lambda)(a-b\lambda) = a^2 - b^2\lambda^2.$$

a. Using the correspondence in Exercise 11, show that $|p|^2 = \det(p)$.
b. Using part a, show that $|pq|^2 = |p|^2|q|^2$.

13. Let $\mathbb{R}[x]$ denote the collection of polynomials in x with coefficients in the real numbers $\mathbb{R}$.
    Show that:
    a. $\mathbb{R}[x]/(x^2+1)$ is isomorphic to the complex numbers.
    b. $\mathbb{R}[x]/x^2$ is isomorphic to the dual real numbers.
    c. $\mathbb{R}[x]/(x^2-1)$ is isomorphic to the hyperbolic numbers.

14. (*Dual Octonions*)
    a   Explain how to define the dual octonions.
    b. Write down a formula for the product of two dual octonions.
    c. What algebraic properties does multiplication have for the dual octonions?

1.2

Algebra

We will begin our study of dual quaternions by investigating some of the unique algebra associated with dual quaternions. We shall focus, in particular, on the notions of conjugates and dot products for dual quaternions. But before we commence with our study of the algebra of dual quaternions, we provide for background, contrast, and completeness a short list of formulas from quaternion algebra that we shall need to invoke from time to time. For further details and derivations for these quaternion formulas and the notion of quaternions as mass-points, see [8,9].

1.2.1 Quaternion Algebra

Notation

- $O = origin\ in\ 3\text{-}dimensions$
- $i, j, k = unit\ vectors\ along\ the\ coordinate\ axes\ in\ 3\text{-}dimensions$
- $v = v_1 i + v_2 j + v_3 k = vector\ in\ 3\text{-}dimensions$
- $p = O + v_1 i + v_2 j + v_3 k = O + v = point\ in\ 3\text{-}dimensions$
- $q = mO + v_1 i + v_2 j + v_3 k = mO + v = mass\text{-}point \leftrightarrow quaternion$
- $mO + v \equiv O + v/m = location\ of\ mass\text{-}point\ in\ 3\text{-}dimensions$

Identity

- $Oq = qO = q$

Products of Basis Vectors

- $i^2 = j^2 = k^2 = -1$
- $ij = k,\ jk = i,\ ki = j,\ ji = -k,\ kj = -i,\ ik = -j$

Product of Vectors

- $v_1 v_2 = -(v_1 \cdot v_2)O + v_1 \times v_2$

DOI: 10.1201/9781003398141-3

Product of Mass-Points (Quaternions)

- $(m_1O + v_1)(m_2O + v_2) = (m_1m_2 - v_1 \cdot v_2)O + m_1v_2 + m_2v_1 + v_1 \times v_2$

Dot Product and Cross Product (Vectors)

- $-(u \cdot v)O = \dfrac{uv + vu}{2}$

- $u \times v = \dfrac{uv - vu}{2}$

Triple Product

- $vuv = (v \cdot v)u - 2(v \cdot u)v$

Conjugate

- $q = mO + v$

- $q^* = mO - v$

- $(pq)^* = q^*p^*$

Length (Norm)

- $|q|^2 = qq^*$

- $|pq| = |p||q|$

Reflection

- $v \mapsto NvN$ reflects vectors v in the plane through the origin O normal to the unit vector N

Rotation

- $q = q(N, \theta/2) = \cos(\theta/2)O + \sin(\theta/2)N$
- $x \mapsto qxq^*$ rotates points and vectors by the angle θ around the line through the origin O parallel to the unit vector N

SLERP (Spherical Linear Interpolation)

- $q_0, q_1 = $ unit quaternions
- $SLERP(q_0, q_1, t) = \dfrac{\sin((1-t)\theta)}{\sin(\theta)}q_0 + \dfrac{\sin(t\theta)}{\sin(\theta)}q_1$

Exercises

1. Show that
    a. $u \times v - (u \cdot v)O = uv$
    b. $u \times v + (u \cdot v)O = -vu$

2. Show that

   a. $q(u \times v)q^{-1} = \left(quq^{-1}\right) \times \left(qvq^{-1}\right)$

   b. $q(u \cdot v)q^{-1} = \left(quq^{-1}\right) \cdot \left(qvq^{-1}\right)$

3. Show that

   c. $(u \times v)^* = -u \times v$

   d. $(u \cdot v)^* = u \cdot v$

1.2.2 Conjugates

There are three kinds of conjugates for dual quaternions: star, bar, and cross. Star is inherited from the standard conjugate for quaternions ($q = wO + xi + yj + zk$, $q^* = wO - xi - yj - zk$), bar mimics complex conjugation ($z = x + yi$, $\bar{z} = x - yi$) but with respect to ε instead of i, and cross incorporates both star and bar.

Conjugates

Star : $(q_1 + q_2\varepsilon)^* = q_1^* + q_2^*\varepsilon$

Bar : $\overline{q_1 + q_2\varepsilon} = q_1 - q_2\varepsilon$

Cross : $q^\dagger = q_1^* - q_2^*\varepsilon$

   Star and bar have the following three properties:

   i. $(pq)^* = q^* p^*$

   ii. $\overline{pq} = \bar{p}\,\bar{q}$

   iii. $\left(\bar{q}\right)^* = \overline{(q^*)}$

The first property follows from the corresponding property for quaternions; the second property is similar to complex conjugation (see Exercise 1). Moreover, we can easily verify the third property, since

$$\overline{(q_1 + q_2\varepsilon)}^* = q_1^* - q_2^*\varepsilon = \overline{(q_1 + q_2\varepsilon)^*}.$$

Therefore

   iv. $q^\dagger = \left(\bar{q}\right)^* = \overline{(q^*)}$.

Hence by properties i–ii, it follows that

   v. $(pq)^\dagger = q^\dagger p^\dagger$

For additional properties of the conjugates star, bar, and cross, see Exercises 1–3.

Exercises

1. Let p, q be two dual quaternions. Show that:

   a. $(pq)^* = q^* p^*$

   b. $\overline{pq} = \overline{p}\,\overline{q}$

   c. $(pq)^\dagger = q^\dagger p^\dagger$

2. Let q be a dual quaternion. Show that:

   a. $(q^*)^* = q$

   b. $\overline{\overline{q}} = q$

   c. $(q^\dagger)^\dagger = q$

3. Let q be a dual quaternion. Show that:

   a. $q^* = q \Rightarrow q$ is a dual real number

   b. $\overline{q} = q \Rightarrow q$ is a quaternion

   c. $q^\dagger = q \Rightarrow q$ is a mass-point (see Section 3.1)

4. Let $q = q_1 + q_2 \varepsilon$ be a dual quaternion. Show that:

   a. $q_1 = \dfrac{q + \overline{q}}{2}$

   b. $q_2 \varepsilon = \dfrac{q - \overline{q}}{2}$

1.2.3 Dot Products and Norms: Lengths and Angles

Dot products can be used to define lengths and angles. In the quaternion algebra dot products, lengths, and angles are real numbers. But in the algebra of dual quaternions dot products, lengths, and angles are dual real numbers.

The dual real numbers are equivalent to the dual quaternions $aO + b\varepsilon$, where a, b are real numbers (see Exercise 1). As usual, we use the symbol O to denote the origin in 3-dimensions, which also serves as the identity for quaternion and dual quaternion multiplication.

To motivate the definition of the dot product for dual quaternions, we begin with a formula for the dot product of two quaternions. Recall that for two quaternions $p = m_p O + v_p$, $q = m_q O + v_q$, by the formulas for conjugates and quaternion multiplication (see Section 2.1)

$$pq^* = (m_p O + v_p)(m_q O - v_q) = (m_p m_q + v_p \cdot v_q)O + (-m_p v_q + m_q v_p - v_p \times v_q)$$

$$qp^* = (m_q O + v_q)(m_p O - v_p) = (m_p m_q + v_p \cdot v_q)O + (m_p v_q - m_q v_p - v_q \times v_p).$$

Therefore, adding these results and dividing by two, we find that for quaternions

$$\frac{pq^* + qp^*}{2} = (m_p m_q + v_p \cdot v_q)O = p \cdot q.$$

Thus, identifying the coefficients of O with the real numbers, we find that the dot product of two quaternions is a real number.

Taking our cue from the quaternions, we define the dot product of two dual quaternions $p = p_1 + p_2\varepsilon$, $q = q_1 + q_2\varepsilon$ by setting

$$p \cdot q = \frac{pq^* + qp^*}{2} = \frac{(p_1 + p_2\varepsilon)(q_1^* + q_2^*\varepsilon) + (q_1 + q_2\varepsilon)(p_1^* + p_2^*\varepsilon)}{2}$$

$$= p_1 \cdot q_1 + (p_1 \cdot q_2 + p_2 \cdot q_1)\varepsilon. \tag{2.1}$$

Notice that with this definition, the dot product for dual quaternions is not a real number; the dot product of two dual quaternions is a dual real number!

Equation (2.1) for the dot product of two dual quaternions mimics Equation (1.1) for the standard product of two dual quaternion with multiplication replaced by dot product. Thus, like multiplication for dual quaternions, this dot product distributes through addition (see Exercise 3). However, unlike multiplication for dual quaternions, this dot product is commutative since

$$p \cdot q = \frac{pq^* + qp^*}{2} = \frac{qp^* + pq^*}{2} = q \cdot p.$$

Even though the dot product of two dual quaternions is a dual real number, nevertheless we can still use this dot product to define when two dual quaternions are orthogonal: two dual quaternions $p \neq q$ are *orthogonal* when the real component of their dot product is zero—that is

$$p \perp q \Leftrightarrow p_1 \cdot q_1 = 0 \Leftrightarrow p_1 \perp q_1 \tag{2.2}$$

In particular, the notions of dot product and orthogonality for quaternions do not change when we consider the quaternions as a subalgebra of the dual quaternions. Hence it follows from Equation (2.2) that in the space of dual quaternions, $O, i, j, k, O\varepsilon, i\varepsilon, j\varepsilon, k\varepsilon$ are mutually orthogonal; thus $O, i, j, k, O\varepsilon, i\varepsilon, j\varepsilon, k\varepsilon$ form an orthogonal basis for the 8-dimensional vector space of dual quaternions over the real numbers (see Exercise 6a). (Notice that since O is the identity, $O\varepsilon = \varepsilon$, but we often use the notation $O\varepsilon$ to emphasize the link with $i\varepsilon, j\varepsilon, k\varepsilon$. Also in the algebra of dual quaternions, the coefficient of ε must be a quaternion and in our notation 1 is not a quaternion.)

We can also define a norm in the space of dual quaternions in the same way we define length in the space of quaternions, by taking the dot product of q with q^*. Thus, we define

$$|q|^2 = q \cdot q^* = qq^* = (q_1 + q_2\varepsilon)(q_1^* + q_2^*\varepsilon)$$

$$= q_1q_1^* + (q_1q_2^* + q_2q_1^*)\varepsilon = |q_1|^2 O + 2(q_1 \cdot q_2)O\varepsilon. \quad (2.3)$$

Hence, the norm for quaternions does not change when we consider the quaternions as a subalgebra of the dual quaternions. Notice again, however, that with this definition the norm of a dual quaternion is not a real number; the norm of a dual quaternion is a dual real number.

Nevertheless, we can still talk about dual quaternions with unit length. In fact, from Equation (2.3) we see that if $q = q_1 + q_2\varepsilon$, then

$$|q|^2 = O \Leftrightarrow |q_1|^2 = O \quad \text{and} \quad q_1 \cdot q_2 = 0. \quad (2.4)$$

In particular:

   i. $|O| = |i| = |j| = |k| = O.$

   ii. $|O\varepsilon| = |i\varepsilon| = |j\varepsilon| = |k\varepsilon| = 0.$

Thus, there exist nonzero dual quaternions with zero length (see Exercise 6c).

Also, by Equation (2.3)

$$|pq|^2 = (pq)(pq)^* = (pq)(q^*p^*) = |p|^2 |q|^2.$$

So just like for quaternions, for dual quaternions

   iii. $|pq| = |p||q|.$ (2.5)

Therefore, the length of a product is equal to the product of the lengths.

Finally,

$$(pq)\cdot(pr) = \frac{(pq)(pr)^* + (pr)(pq)^*}{2} = \frac{(pq)(r^*p^*) + (pr)(q^*p^*)}{2}$$

$$= \frac{p(qr^* + rq^*)p^*}{2} = p(q\cdot r)p^*,$$

so for dual quaternions since the dot product is a dual real number

iv. $(pq)\cdot(pr) = |p|^2(q\cdot r).$ \hfill (2.6)

Hence by Equations (2.5) and (2.6), multiplication with unit norm dual quaternions preserves both length and dot product.

Trigonometric functions can also be defined using the dot product. We define the cosine of the angle $\theta_{q,r}$ between two dual quaternions q, r in the usual way by setting

$$\cos(\theta_{q,r}) = \frac{q\cdot r}{|q||r|} \qquad \text{provided that } |q||r| \neq 0 \hfill (2.7)$$

Notice yet again that with this definition, the cosine of the angle between two dual quaternions is not a real number; the cosine, like the dot product and the norm, is a dual real number. Nevertheless, we have the following consequences for the cosine between two dual quaternions:

v. $q \perp r \Leftrightarrow real\ part\left(\cos(\theta_{q,r})\right) = 0 \Leftrightarrow q_1 \cdot r_1 = 0$ by Equations (2.2) and (2.7);

vi. $\cos(\theta_{pq,\ pr}) = \dfrac{(pq)\cdot(pr)}{|pq||pr|} = \dfrac{q\cdot r}{|q||r|} = \cos(\theta_{q,r})$ for $|p||q||r| \neq 0$ by

Equations (2.5)–(2.7).

Thus, multiplication by a nonzero dual quaternion does not change the cosine of the angle between two dual quaternions. Therefore, multiplication by unit dual quaternions preserves both lengths and angles.

Exercises

1. Show that the set $\{aO + b\varepsilon \mid a, b \text{ are real numbers}\}$ is a commutative subalgebra of the dual quaternions isomorphic to the dual real numbers.

2. Let $q = q_1 + q_2\varepsilon$ be a dual quaternion and let $d = aO + b\varepsilon$ be a dual real number. Show that:

   a. $|d|^2 = dd^* = d^2 = a^2O + 2ab\varepsilon$

   b. $|dq|^2 = d^2|q|^2$.

3. Consider three dual quaternions $p = p_1 + p_2\varepsilon$, $q = q_1 + q_2\varepsilon$, $r = r_1 + r_2\varepsilon$. Show that:

   a. $p \cdot (q + r) = p \cdot q + p \cdot r$

   b. $(p + q) \cdot r = p \cdot r + q \cdot r$

   c. Conclude that for dual quaternions the dot product distributes through addition.

4. Let $q = q_1 + q_2\varepsilon$ be a dual quaternion. Show that $q \cdot \bar{q} = q_1 \cdot q_1$.

5. Let p, q be dual quaternions. Show that:

   a. $(p \cdot q)^* = p \cdot q$.

   b. $(pq) \cdot (pq) = |p|^2|q|^2$.

6. Show that:

   a. $O, i, j, k, O\varepsilon, i\varepsilon, j\varepsilon, k\varepsilon$ are mutually orthogonal.

   b. $|O| = |i| = |j| = |k| = O$.

   c. $|O\varepsilon| = |i\varepsilon| = |j\varepsilon| = |k\varepsilon| = 0$.

7. Let $p, q, r.s$ be dual quaternions. Show that $p \perp r \Leftrightarrow (p + q\varepsilon) \perp (r + s\varepsilon)$.

1.3

Geometry

Dual quaternions are vectors in 8-dimensions with the canonical basis: $O, i, j, k, O\varepsilon, i\varepsilon, j\varepsilon, k\varepsilon$. But we are interested in investigating geometry in 3-dimensions. Therefore, we need a way to represent points, vectors, lines, and planes in 3-dimensions inside the 8-dimensional vector space of dual quaternions.

1.3.1 Points and Vectors in the Space of Dual Quaternions

When we study quaternions, we use the 4-dimensional vector space generated by the basis O, i, j, k to represent mass-points and vectors in 3-dimensions: three of the dimensions are spatial, the fourth dimension—the dimension along O—is due to the mass. The quaternions are a subspace of the space of dual quaternions, so we could adopt this subspace to represent mass-points and vectors using dual quaternions. But the quaternions are not a subalgebra of the dual quaternions relative to the coefficients of dual real numbers. Moreover, adopting this approach will not lead to any interesting new geometry. Rather points and vectors in 3-dimensions are represented somewhat differently using dual quaternions. We summarize these different representations in Table 3.1.

TABLE 3.1

Representations for Points and Vectors Using Quaternions and Dual Quaternions

	Quaternions	Dual Quaternions
Points	$P = O + p_1 i + p_2 j + p_3 k$	$P = O + (p_1 i + p_2 j + p_3 k)\varepsilon$
	$P \leftrightarrow O + v_P$	$P \leftrightarrow O + v_P\varepsilon$
Vectors	$v = v_1 i + v_2 j + v_3 k$	$v = (v_1 i + v_2 j + v_3 k)\varepsilon$
	$v \leftrightarrow v$	$v \leftrightarrow v\varepsilon$

Notice that the four basis vectors $O, i\varepsilon, j\varepsilon, k\varepsilon$ are invoked to represent points and vectors using dual quaternions. The symbol O represents the origin in 3-dimensions; the origin O also represents the identity for quaternion and dual quaternion multiplication. Mass-points are represented

DOI: 10.1201/9781003398141-4

by expression of the form $P = mO + (p_1 i + p_2 j + p_3 k)\varepsilon = mO + v_P \varepsilon$. The point $P = O + (p_1 i + p_2 j + p_3 k)\varepsilon$ is located at the position with coordinates (p_1, p_2, p_3). The mass-point $P = mO + (p_1 i + p_2 j + p_3 k)\varepsilon = mO + v_P \varepsilon$ is located at the position with coordinates $\left(\dfrac{p_1}{m}, \dfrac{p_2}{m}, \dfrac{p_3}{m}\right)$. We shall write $mO + v_P \varepsilon \equiv O + \dfrac{v_P}{m}\varepsilon$ to indicate that the mass-point $mO + v_P \varepsilon$ is located at the point $O + \dfrac{v_P}{m}\varepsilon$. In Section 3.2 we shall see that the remaining basis vectors $O\varepsilon, i, j, k$ are used to represent planes in the space of dual quaternions.

Exercises

1. Let $P = O + v_P \varepsilon$ represent a point in the space of dual quaternions. Describe the locations of the points represented by the three conjugates: $\bar{P}, P^*, P^\dagger$.

2. Let P, Q be the dual quaternion representations for two points. Show that:

   a. $|P| = O$.

   b. $P \cdot Q = O$.

3. Show that the set of mass-points

   $\{mO + v\varepsilon \mid m \text{ is a dual real number and } v \text{ is a vector in 3-dimensions}\}$

   is a commutative subalgebra of the dual quaternions relative to the dual real numbers.

4. Let $P = O + v_P \varepsilon$ and $Q = O + v_Q \varepsilon$ be the dual quaternion representations for two points.

   a. Show that the mass-point $P + Q$ is located at the midpoint of the line segment from P to Q.

   b. Describe the location of the point PQ.

1.3.2 Planes in the Space of Dual Quaternions

The basis vectors $O\varepsilon, i, j, k$ not used to represent points and vectors are used to represent planes in the space of dual quaternions. In Table 3.2 we compare the representations for planes in the spaces of quaternions and dual quaternions.

In the space of quaternions a plane π is represented by a normal vector $v_\pi = ai + bj + ck$ and its signed distance $d/|v_\pi|$ from the origin O along this

TABLE 3.2

Representations for Planes Using Quaternions and
Dual Quaternions

	Quaternions	Dual Quaternions
Planes	$\pi \equiv ax + by + cz + d = 0$	$\pi \equiv ax + by + cz + d = 0$
	$\pi \leftrightarrow ai + bj + ck + dO$	$\pi \leftrightarrow ai + bj + ck + dO\varepsilon$
	$\pi \leftrightarrow dO + v_\pi$	$\pi \leftrightarrow dO\varepsilon + v_\pi$

normal vector. Similarly, in the space of dual quaternions, v_π represents a normal vector to the plane π, and $d/|v_\pi|$ is the signed distance along this normal of the plane π from the origin O (see Exercise 2b).

Notice, however, that quaternion representations do not distinguish clearly between mass-points and planes nor between ordinary vectors and normal vectors. Dual quaternions, in contrast, do distinguish clearly between points and planes and between standard vectors and normal vectors by using different basis elements to represent these different geometric entities. *This ability to distinguish clearly between points and planes and between ordinary vectors and normal vectors is one of the many advantages of dual quaternions over quaternions.*

In the dual quaternion representation for a plane $\pi = v_\pi + dO\varepsilon$, the sign of the normal vector v_π is ambiguous because π and $-\pi$ represent the same plane. But for planes not passing through the origin the sign of the product dv_π is not ambiguous. The normal vector dv_π when attached to the point on the plane closest to the origin always points toward the origin (see Exercise 3a). Moreover, $-\dfrac{dv_\pi}{|v_\pi|^2}$ is the vector from the origin to the point on the plane closest to the origin (see Exercise 3b). Similar results hold, of course, for the quaternion representation for planes.

With these representations for points and planes in the space of dual quaternions, we can use the dot product to determine when a point P lies on a plane π.

Proposition 3.1

A point P lies on the plane π if and only if $P \cdot \pi = 0$.

Proof: Let $P = O + (x'i + y'j + z'k)\varepsilon$ and $\pi = ai + bj + ck + dO\varepsilon$. Then the plane π corresponds to the points whose coordinates satisfy the equation $ax + by + cz + d = 0$. But

$$P \cdot \pi = (ax' + by' + cz' + d)O\varepsilon$$

Thus, the point P lies on the plane π if and only if $P \cdot \pi = 0$. ◊

Corollary 3.2

Let P be a point and let v be a normal vector in the space of dual quaternions. Then $\pi = v - P \cdot v$ represents the plane passing through the point P with normal vector v.

Proof: Let $P = O + u\varepsilon$. Then by Equation (2.1)

i. $P \cdot v = (O + u\varepsilon) \cdot v = (u \cdot v)O\varepsilon$

ii. $P \cdot (P \cdot v) = (O + u\varepsilon) \cdot (u \cdot v)O\varepsilon = (u \cdot v)O\varepsilon$

Thus, by (i), π represents a plane with normal vector v because v represents a normal vector and $P \cdot v$ is a scalar multiple of $O\varepsilon$. Also, subtracting (ii) from (i) yields

$$P \cdot \pi = P \cdot v - P \cdot (P \cdot v) = 0.$$

Therefore, by Proposition 3.1, the point P lies on the plane π. ◊

Exercises

1. Let $\pi = v_\pi + d_\pi O\varepsilon$ and $\rho = v_\rho + d_\rho O\varepsilon$ be dual quaternion representations for two planes. Show that:
   a. $|\pi|^2 = |v_\pi|^2$.
   b. $\pi \cdot \rho = v_\pi \cdot v_\rho$.
2. Show that in the dual quaternion representation for a plane $\pi = ai + bj + ck + dO\varepsilon$
   a. $v_\pi = ai + bj + ck$ represents a vector normal to the plane π.
   b. $d/|v_\pi|$ is the signed distance along the normal of the plane π from the origin O.
3. Consider a plane $\pi = v_\pi + dO\varepsilon$ with normal vector v_π and $d \neq 0$. Show that:
   a. the normal vector dv_π when attached to the point on the plane closest to the origin always points toward the origin.
   b. $-\dfrac{dv_\pi}{|v_\pi|^2}$ represents the vector from the origin to the point on the plane closest to the origin.
4. Let v be a vector and let π be a plane through the origin. Show that the vector v lies on the plane π if and only if $v \cdot \pi = 0$.

5. Show that the point

$$P = O + \frac{(v_1 \cdot C_1)(v_2 \times v_3) + (v_2 \cdot C_2)(v_3 \times v_1) + (v_3 \cdot C_3)(v_1 \times v_2)}{det(v_1 \ v_2 \ v_3)}$$ is the

intersection point of the three planes $\pi_k = v_k - C_k \cdot v_k$, $k = 1, 2, 3$.

1.3.3 Lines in the Space of Dual Quaternions

Lines in 3-dimensions can be represented in at least five different ways:

1. Two points determine a line. (Join of two points)
2. One point and one vector determine a line. (Join of a point and a vector)
3. Two planes determine a line. (Intersection of two planes)
4. A vector parallel to the line and a normal vector perpendicular to the plane containing both the line and the origin. (Plucker Coordinates)
5. A vector parallel to the line and a plane containing both the line and the origin. (Dual Plucker Coordinates)

Each of these representations requires six parameters. The first three representations are quite well-known, but the two Plucker coordinate methods are perhaps less familiar. Nevertheless, it is these Plucker coordinate representations that are most natural in the space of dual quaternions. Therefore, for the uninitiated, we shall now review both Plucker coordinates and dual Plucker coordinates.

(Warning: The following two subsections make substantial use of cross products. For readers unfamiliar with the algebraic and geometric properties of the cross product, see the Appendix for a list of identities. For the proofs of these identities, see Part II, Chapter 1, Section 4.1 and the Exercises therein.)

1.3.3.1 Plucker Coordinates

Consider a line l through a point P and parallel to a vector v. Let $u = (P - O) \times v$. The six *Plucker coordinates* of the line l consist of the pair $(v, u) = (v, (P - O) \times v)$. The component v is a vector parallel to the line l, and the component u is a vector normal to the plane containing both the line l and the origin O. Crucially, the component $u = (P - O) \times v$ is independent of the choice of the point P on the line l because if Q is any other point on the line l, then $P - Q$ is parallel to v. Therefore

$$\big((P-O)-(Q-O)\big)\times v=(P-Q)\times v=0$$

$$(P-O)\times v=(Q-O)\times v.$$

Given the Plucker coordinates (v, u) of a line l, we can find a point Q on the line l in the following manner. If the line l passes through the origin O, then $u = 0$, which indicates that $Q = O$ is a point on the line l. Otherwise, let Q be the point on the line l closest to the origin O. Then the vector $Q - O$ is perpendicular to the line vector v. Moreover since $u = (Q - O) \times v$ is a cross product, u is perpendicular to both $Q - O$ and v. Thus $Q - O$, v, u are mutually orthogonal vectors. Therefore, we can recover $Q - O$ from the Plucker coordinates (v, u) of the line l using cross products, since

$$Q-O=\frac{v\times\big((Q-O)\times v\big)}{|v|^{2}}=\frac{v\times u}{|v|^{2}}.$$

Hence, in the space of dual quaternions

$$Q=O+\frac{v\times u}{|v|^{2}}\varepsilon. \tag{3.1}$$

Once we have one point Q on the line l, we can find arbitrarily many points P on the line l, using the formula

$$P=Q+\lambda v\varepsilon, \text{ where } \lambda \text{ is a real number.}$$

We can express the Plucker coordinates (v, u) of a line l in a single expression by writing

$$l=v+u\varepsilon=v+(P-O)\times v$$

$$v\cdot u=v\cdot\big((P-O)\times v\big)=0.$$

We started by saying that the Plucker coordinates of a line l consist of a vector v parallel to the line and a normal vector u perpendicular to the plane containing both the line and the origin. But in the sum $l = v + u\varepsilon$ the symbol v represents a plane through the origin with normal vector parallel to the line l, and $u\varepsilon = (P - Q) \times v$ represents a vector in this plane. It may seem strange that in this representation v is a normal vector parallel to the line l rather than a vector parallel to the line l, but the expression $P - Q$ is already a vector—a multiple of ε—so in order to distinguish clearly between the two components of the Plucker coordinates, v cannot also be multiplied by ε. Hence we write $l = v + u\varepsilon$ rather than $l = u + v\varepsilon$.

Now any expression of the form

$$l = v + u\varepsilon \text{ with } v \cdot u = 0$$

represents the line l in the space of dual quaternions in terms of the Plucker coordinates of the line l. The normal vector v is parallel to the direction of the line l, the point $Q = O + \dfrac{v \times u}{|v|^2}\varepsilon$ is a point on the line l, and since $v \cdot u = 0$,

$$u\varepsilon = \left(\frac{v \times u}{|v|^2}\varepsilon\right) \times v = (Q - O) \times v.$$

In retrospect it may seem odd that Plucker coordinates completely specify a line. After all, the Plucker coordinates consist of two orthogonal vectors (v, u), so the pair (v, u) seems to contain information only about directions but not about locations. However, these vectors contain more than just directional information: the ratio $|u|/|v|$ of their lengths is the distance $|Q - O|$ from the origin O to the line l (see Exercise 1). Here then is where information about the location of the line l is stored. The magnitude of the vector v is arbitrary, but the magnitude the ratio $|u|/|v|$ is an invariant.

Exercises

1. Let $l = v + u\varepsilon$ with $v \cdot u = 0$ represent a line. Show that the ratio $|u|/|v|$ is the distance from the origin O to the line l independent of the choice of the length of the vector v.

2. Let $l_1 = v_1 + u_1\varepsilon$ and $l_2 = v_2 + u_2\varepsilon$ represent lines in terms of their Plucker coordinates. Show that l_1 and l_2 represent the same line if and only if $l_2 = cl_1$ for some constant $c \neq 0$.

3. Show that up to a constant multiple, the Plucker coordinates representing the line l are independent of the choice of the point P lying on l and the vector v parallel to l.

4. Show that the point P lies on the line $l = v + u\varepsilon$ if and only if $(P - O) \times v = u\varepsilon$.

5. Consider a line $l = v + u\varepsilon$ and a plane $\pi = dO\varepsilon + v_\pi$. Let p_l be a point on the line l. Show that the following four statements are equivalent:

   a. the line l lies on the plane π.

   b. $v \cdot v_\pi = 0$ and $d = \dfrac{\det(u\,v\,v_\pi)}{|v|^2}$.

   c. $v \cdot \pi = 0$ and $p_l \cdot \pi = 0$.

   d. $(v + p_l) \cdot \pi = 0$.

6. Show that the line $l = v + u\varepsilon$ intersects the plane $\pi = dO\varepsilon + v_\pi$ at the point $P = O + \dfrac{v \times u}{|v|^2}\varepsilon + \lambda v\varepsilon$, where $\lambda = -\dfrac{d\,|v|^2 + \det(v\ u\ v_\pi)}{(v \cdot v_\pi)\,|v|^2}$, provided that $v \cdot v_\pi \neq 0$.

7. Let P, Q be two points on the line l. Show that the Plucker coordinates of l are

$$\big((Q - P), (P - O) \times (Q - O)\big).$$

8. Let $l_1 = v_1 + u_1\varepsilon$ and $l_2 = v_2 + u_2\varepsilon$ represent two lines in terms of their Plucker coordinates. Let $Q_1 = O + \dfrac{v_1 \times u_1}{|v_1|^2}\varepsilon$, $Q_2 = O + \dfrac{v_2 \times u_2}{|v_2|^2}\varepsilon$, and set $\pi = v_1 \times v_2 - Q_1 \cdot (v_1 \times v_2)$. Show that:

   a. The line l_1 lies on the plane π.
   b. The line l_2 is parallel to the plane π.
   c. The lines l_1 and l_2 are skew if and only if

   $$Q_2 \cdot \pi = \det(v_1\ v_2\ Q_2 - Q_1) \neq 0.$$

9. Let $l_1 = v_1 + u_1\varepsilon$ and $l_2 = v_2 + u_2\varepsilon$ represent two intersecting (non-skew, nonparallel) lines in terms of their Plucker coordinates. Let

$$w_1 = \frac{v_1 \times u_1}{|v_1|^2},\quad w_2 = \frac{v_2 \times u_2}{|v_2|^2},\quad \text{and set}\quad \lambda = \frac{\big((w_2 - w_1) \times v_2\big) \cdot (v_2 \times v_1)}{|v_2 \times v_1|^2}.$$

Show that the lines l_1, l_2 intersect at the point $P = O + (w_2 + \lambda v_2)\varepsilon$.

1.3.3.2 Dual Plucker Coordinates

Consider a line l that is the intersection of two planes $\pi_1 = v_1 + d_1 O\varepsilon$ and $\pi_2 = v_2 + d_2 O\varepsilon$. The six *dual Plucker coordinates* (v, u) of the line l are defined by setting

$$v = v_1 \times v_2$$

$$u = d_1\pi_2 - d_2\pi_1 = d_1 v_2 - d_2 v_1.$$

The parameters v and u have the following geometric interpretation: v is parallel to the direction of the line l, since v is perpendicular to the normal vectors v_1, v_2 of the two planes π_1, π_2; u is the plane containing both the origin O and the line l, since the origin lies on u and any point on both planes π_1, π_2 necessarily lies on u (see Exercise 1). Since v is perpendicular to the normal vectors v_1, v_2, we have the additional constraint

$$v \cdot u = 0.$$

Given the dual Plucker coordinates (v, u) of a line l, we can find a point Q on the line l in the following manner. If the line l passes through the origin O, then $d_1 = d_2 = 0$ so $u = 0$, which indicates that $Q = O$ is a point on the line l. Otherwise let

$$Q = O + \frac{v \times u}{|v|^2} \varepsilon. \tag{3.2}$$

Then it is straightforward to check that

$$Q \cdot \pi_1 = Q \cdot \pi_2 = 0. \quad \text{(see Exercise 2a)}.$$

Thus, the point Q lies on both π_1 and π_2 and therefore on the intersection line l of the two planes π_1 and π_2. Notice that this formula for the point Q on the line l using dual Plucker coordinates is identical to the formula for the point Q on the line l using Plucker coordinates. Once we have one point Q on the line l, we can find arbitrarily many points P on the line l, using the formula

$$P = Q + \lambda v \varepsilon, \text{ where } \lambda \text{ is a real number}.$$

We can express the dual Plucker coordinates (v, u) of a line l in a single expression by writing

$$l = v + u \varepsilon = v_1 \times v_2 + (d_1 v_2 - d_2 v_1) \varepsilon$$

$$v \cdot u = v_1 \times v_2 \cdot (d_1 v_2 - d_2 v_1) = 0.$$

We started by saying that the dual Plucker coordinates of a line l consist of a vector v parallel to the line and a normal vector u perpendicular to the plane containing both the line and the origin. But in this sum v represents a plane through the origin with normal vector parallel to the line l, and $u \varepsilon$ represents a vector in this plane rather than a plane. We seem to have interchanged vectors and normal vectors. We have made this choice, however, so that this representation for dual Plucker coordinates is consistent with our representation for Plucker coordinates. Hence, we write $l = v + u \varepsilon$ rather than $l = u + v \varepsilon$.

Now any expression of the form

$$l = v + u \varepsilon \text{ with } v \cdot u = 0$$

represents the line l in the space of dual quaternions in terms of the dual Plucker coordinates of the line. We can recover two planes π_1, π_2 whose intersection is the line l from this representation in the following manner. If $u = 0$, then since π_1 and π_1 are necessarily distinct, it follows that $d_1 = d_2 = 0$. Hence we need only find two normal vectors whose cross product is v. Let w be any unit normal vector perpendicular to v. Then $v, w, v \times w$ are mutually orthogonal vectors, so $w \times (v \times w) = v$. Thus when $u = 0$, we can simply choose $\pi_1 = w$ and $\pi_2 = v \times w$—that is, we can simply choose two appropriate planes passing through the origin.

It remains to treat the case where $u \neq 0$. Now we need to find two planes

$$\pi_1 = v_1 + d_1 O \varepsilon$$

$$\pi_2 = v_2 + d_2 O \varepsilon$$

such that

$$u = d_1 v_2 - d_2 v_1$$

$$v = v_1 \times v_2$$

Therefore, we need only find normal vectors w_1, w_2 such that

$$d_1 w_2 - d_2 w_1 = u$$

$$w_1 \times w_2 = v.$$

Then we can simply set

$$\pi_1 = w_1 + d_1 O \varepsilon$$

$$\pi_2 = w_2 + d_2 O \varepsilon.$$

Lemma 3.3

Suppose that $u, v \neq 0$ and that $v \cdot u = 0$. Set

$$w_1 = \alpha \frac{u}{|u|} - d_1 \frac{|v|}{|u|} \left(\frac{v \times u}{|v||u|} \right)$$

$$w_2 = \gamma \frac{u}{|u|} - d_2 \frac{|v|}{|u|} \left(\frac{v \times u}{|v||u|} \right)$$

where α, γ are constants that satisfy the constraint $d_1 \gamma - d_2 \alpha = |u|$. Then

i. $d_1 w_2 - d_2 w_1 = u$

ii. $w_1 \times w_2 = v$.

Proof: Part i is immediate by direct computation. To prove part ii, observe that since $v \cdot u = 0$, the vectors $\dfrac{v}{|v|}, \dfrac{u}{|u|}, \dfrac{v \times u}{|v||u|}$ form an orthonormal basis. Therefore

$$\frac{u}{|u|} \times \frac{v \times u}{|v||u|} = \frac{v}{|v|}.$$

Hence since $d_1 \gamma - d_2 \alpha = |u|$,

$$w_1 \times w_2 = \left(\alpha \frac{u}{|u|} - d_1 \frac{|v|}{|u|} \left(\frac{v \times u}{|v||u|} \right) \right) \times \left(\gamma \frac{u}{|u|} - d_2 \frac{|v|}{|u|} \left(\frac{v \times u}{|v||u|} \right) \right)$$

$$= \frac{|v|}{|u|} \underbrace{(d_1 \gamma - d_2 \alpha)}_{|u|} \underbrace{\left(\frac{u}{|u|} \times \frac{v \times u}{|v||u|} \right)}_{\frac{v}{|v|}} = v. \ \Diamond$$

Exercises

1. Let (v, u) be the dual Plucker coordinates of the line l. Show that u represents the plane containing both the origin O and the line l.

2. Consider the line l that is the intersection of the two planes $\pi_1 = v_1 + d_1 O\varepsilon$ and $\pi_2 = v_2 + d_2 O\varepsilon$. Let $v = v_1 \times v_2$ and $u = d_1 v_2 - d_2 v_1$ be the dual Plucker coordinates of the line l, and let $Q = O + \dfrac{v \times u}{|v|^2} \varepsilon$.

   Show that:

   a. $Q \cdot \pi_1 = Q \cdot \pi_2 = 0$.

   b. Q is a point on the intersection line of the two planes π_1 and π_2.

   c. $Q - O$ is orthogonal to the line vector v.

   d. Q is the point on the line l closest to the origin O.

   e. $\dfrac{|u|}{|v|} = |Q - O| =$ the distance from the origin O to the line l.

      (Hint: $v \cdot u = 0$.)

   f. $(Q - O) \times v = u$

   g. Dual Plucker coordinates and Plucker coordinates for the same line l are equal up to constant multiples.

3. Let $l_1 = v_1 + u_1$ and $l_2 = v_2 + u_2\varepsilon$ represent lines in terms of their dual Plucker coordinates. Show that l_1 and l_2 represent the same line if and only if $l_2 = cl_1$ for some constant $c \neq 0$.

4. Show that, up to a constant multiple, the dual Plucker coordinates representing the line l are independent of the choice of the two planes whose intersection is the line l.

1.3.4 Duality in the Space of Dual Quaternions

In 3-dimensions we sometimes use normal vectors to represent planes through the origin. In this sense, *vectors are dual to planes in 3-dimensions*. We shall now formalize this notion of duality in the algebra of dual quaternions by defining duality for the basis vectors and then extending to arbitrary dual quaternions by linearity.

We adopt the notation $s = r^\#$ to indicate that s is dual to r. If the vector s is dual to the plane r, then the plane r is also dual to the vector s, so we define duality for the basis elements in dual pairs:

$$i^\# = i\varepsilon \text{ and } (i\varepsilon)^\# = i \tag{3.3}$$

$$j^\# = j\varepsilon \text{ and } (j\varepsilon)^\# = j \tag{3.4}$$

$$k^\# = k\varepsilon \text{ and } (k\varepsilon)^\# = k \tag{3.5}$$

$$O^\# = O\varepsilon \text{ and } (O\varepsilon)^\# = O \tag{3.6}$$

Thus, the unit vector $i\varepsilon$ along the x-axis and the yz-plane represented by i ($x = 0$) are dual to each other. Similarly, the unit vector $j\varepsilon$ along the y-axis and the xz-plane represented by j ($y = 0$) are dual to each other, and the unit vector $k\varepsilon$ along the z-axis and the xy-plane represented by k ($z = 0$) are dual to each other. Hence the coordinate planes and their unit normal vectors are dual. To complete this definition of duality, the origin O and the plane at infinity $O\varepsilon$ are dual. Thus, the plane at infinity is dual to a point, not a vector, since there is not one unique direction normal to the plane at infinity; in some sense all directions emanating from the origin are normal to the plane at infinity. Therefore, we choose the origin to represent the dual to the plane at infinity, and the plane at infinity to be dual to the point at the origin. Hence for the basis elements, duality is defined so that

$$r \cdot r^\# = \varepsilon \tag{3.7}$$

By Equations (3.3)–(3.6) we can transform between the basis elements and their duals either by removing or by inserting a factor of ε. Thus, in general, if $q = q_1 + q_2\varepsilon$, then by linearity $q^{\#} = q_2 + q_1\varepsilon$. Therefore for all dual quaternions

$$(q^{\#})^{\#} = q. \tag{3.8}$$

Moreover, we have the following table for duality between mass-points and planes, vectors and normal vectors, and lines and dual lines. Notice in particular that duality converts distance into mass. This observation will be crucial when we investigate perspective and pseudo-perspective projections in Section 1.7 (Table 3.3).

TABLE 3.3

Duality for Mass-Points, Vectors, Planes, and Lines

	Primal Representation	Dual Representation
Mass-Points	$P = mO + (p_1 i + p_2 j + p_3 k)\varepsilon$	$P^{\#} = mO\varepsilon + (p_1 i + p_2 j + p_3 k)$
	$P \leftrightarrow mO + v_P\varepsilon$	$P^{\#} \leftrightarrow mO\varepsilon + v_P$
Vectors	$v = (v_1 i + v_2 j + v_3 k)\varepsilon$	$v^{\#} = v_1 i + v_2 j + v_3 k$
	$v \leftrightarrow v\varepsilon$	$v^{\#} \leftrightarrow v$
Planes	$\pi = ai + bj + ck + dO\varepsilon$	$\pi^{\#} = (ai + bj + ck)\varepsilon + dO$
	$\pi \leftrightarrow v_{\pi} + dO\varepsilon$	$\pi^{\#} \leftrightarrow v_{\pi}\varepsilon + dO$
Lines	$l = (v_1 i + v_2 j + v_3 k) + (u_1 i + u_2 j + u_3 k)\varepsilon$	$l^{\#} = (u_1 i + u_2 j + u_3 k) + (v_1 i + v_2 j + v_3 k)\varepsilon$
	$l = v + u\varepsilon$	$l^{\#} \leftrightarrow u + v\varepsilon$

Exercises

1. Suppose that p represents a point and q represents a vector in the space of dual quaternions. Show that it is *not* necessarily true that
   a. $|p^{\#}| = 1$
   b. $|q^{\#}| = 0$

2. Show that the set

$$\left\{(mO + v\varepsilon)^{\#} \mid m \text{ is a dual real number and } v \text{ is a vector in 3-dimensions}\right\}$$

   is a *not* a subalgebra of the algebra of dual quaternions. Compare to Section 1.3.1, Exercise 3.

3. Let P be a mass-point and let π be a plane in the space of dual quaternions.

a. Show that $P^{\#} \cdot \pi^{\#} = P \cdot \pi$

b. Conclude from part a that a mass-point P lies on a plane π if and only if $P^{\#} \cdot \pi^{\#} = 0$.

c. Conclude from part b that a mass-point P lies on a plane π if and only if the dual mass-point $\pi^{\#}$ lies on the dual plane $P^{\#}$.

4. Show that if the mass-point $p = mO + mv_p\varepsilon$ lies on the line $l = v + u\varepsilon$, then the dual line $l^{\#} = u + v\varepsilon$ lies on the dual plane $p^{\#} = mO\varepsilon + mv_p$. (Hint: See Section 1.3.3.1, Exercises 4 and 5.)

5. Let v be a vector. Show that $\left(v \cdot v^{\#}\right)^{\#} = |v|^2$.

6. Let P be a point and let π be a plane with a unit normal vector. Show that $dist(P,\pi) = (P \cdot \pi)^{\#}$.

7. Let P be a point and let $l = v + u\varepsilon$ with $v \cdot u = 0$ be a line. Show that

$$dist(P,l) = \left| \left(\frac{(P-O) \times v - u\varepsilon}{|v|} \right)^{\#} \right|.$$

8. Let $l = v + u\varepsilon$ with $v \cdot u = 0$ represent a line. How is $l^{\#} = u + v\varepsilon$ related geometrically to l?

9. Let $q = q_1 + q_2\varepsilon$ be a dual quaternion. Show that:

a. $q_1 = (q\varepsilon)^{\#}$

b. $q_2\varepsilon = q^{\#}\varepsilon$

Compare to Section 1.2.2, Exercise 4.

10. Consider two points P, Q. Let

$$v_p = (P - O)^{\#}$$

$$v_q = (Q - O)^{\#}.$$

Show that the Plucker coordinates of the line l joining the points P and Q are given by $l = \left(v_q - v_p\right) + \left(v_p \times v_q\right)\varepsilon$

1.4

Rigid Motions

Rigid motions are at the heart of Euclidean geometry. The rigid motions consist of composites of rotations, translations, and reflections. Using the quaternion algebra, we can compute rotations and reflections, but not translations. *One of the main advantages of dual quaternions is that we can also compute translations in the algebra of dual quaternions.*

1.4.1 Rotation and Translation

Rotation in the quaternion algebra can be computed by sandwiching a point or a vector between a unit quaternion and its conjugate [9, Chapter 6]. Rotation using sandwiching with unit quaternions also works with the representation for points and vectors in the space of dual quaternions. Indeed, let N be a unit vector, and set

$$q = q(N, \theta/2) = \cos(\theta/2)O + \sin(\theta/2)N.$$

Then in the quaternion algebra the sandwiching maps

$$v \mapsto qvq^*$$

$$P \mapsto qPq^*$$

rotate vectors and points by the angle θ around the line through the origin O parallel to the unit vector N. Moreover, since ε commutes with every quaternion q,

$$q(v\varepsilon)q^* = \left(qvq^*\right)\varepsilon$$

$$qPq^* = q(O + v_P\varepsilon)q^* = qOq^* + q(v_P\varepsilon)q^* = O + \left(qv_Pq^*\right)\varepsilon \tag{4.1}$$

Thus, just like for the quaternion representation of points and vectors, sandwiching between the unit quaternion q and its conjugate q^* also rotates points and vectors in the dual quaternion representation by the

DOI: 10.1201/9781003398141-5

angle θ around the line through the origin O parallel to the unit vector N. Nothing much really changes for rotation when the vectors are multiplied by ε. Notice, however, that in the space of dual quaternions N represents a a normal vector perpendicular to the plane of rotation, rather than a vector parallel to the axis of rotation. Consequently, the geometric interpretation of the quaternion N changes, but the formula for rotation remains the same.

The utility of the novel representation for points and vectors using dual quaternions becomes clear, however, when we multiply two points P,Q. For quaternions

$$PQ = (O + v_P)(O + v_Q) = O + v_P + v_Q + v_P v_Q = (P + v_Q) + v_P v_Q.$$

But for dual quaternions

$$PQ = (O + v_P \varepsilon)(O + v_Q \varepsilon) = O + (v_P + v_Q)\varepsilon + v_P v_Q \varepsilon^2$$

$$= (O + v_P \varepsilon) + v_Q \varepsilon = P + v_Q \varepsilon.$$

Thus, for dual quaternions, unlike for quaternions, multiplying the point P by the point Q translates the point P by the vector $v_Q \varepsilon$; unlike quaternion multiplication for points, there is no extraneous $v_P v_Q$ term because $\varepsilon^2 = 0$. In this way dual quaternions support translation. Since the dual quaternions contain the quaternions as a subalgebra—the subalgebra where the coefficient of ε is zero—both translations and rotations can be computed using dual quaternions.

However, we would like to represent translation and rotation in compatible ways so that we can compose these transformations by multiplication in the space of dual quaternions. Rotation is represented by sandwiching a point P between a unit quaternion q and its conjugate q^*—that is, by the map $P \mapsto qPq^*$. Therefore, we would also like to represent translation by sandwiching a point between two dual quaternions.

Consider then the following sandwiching formula for translation. Let $t\varepsilon$ be a unit vector and set

$$T = T(t,d/2) = O + \frac{d}{2}t\varepsilon.$$

Now for any point $P = O + v_P \varepsilon$:

$$TPT = \left(O + \frac{d}{2}t\varepsilon\right)(O + v_P \varepsilon)\left(O + \frac{d}{2}t\varepsilon\right) = \left(O + (v_P + \frac{d}{2}t)\varepsilon\right)\left(O + \frac{d}{2}t\varepsilon\right)$$

$$= O + (v_P + dt)\varepsilon = (O + v_P \varepsilon) + dt\varepsilon = P + dt\varepsilon. \tag{4.2}$$

In this way we can indeed use sandwiching—that is, the map $P \mapsto TPT$—to translate a point P in the direction of the unit vector $t\varepsilon$ by the distance d. Similarly, for any vector $v\varepsilon$—

$$T(v\varepsilon)T = \left(O + \frac{d}{2}t\varepsilon\right)(v\varepsilon)\left(O + \frac{d}{2}t\varepsilon\right) = v\varepsilon$$

that is, vectors are unaffected by translation—so this sandwiching map for translation behaves correctly on vectors as well as on points. Notice that we need the factor $1/2$ to multiply the translation vector $dt\varepsilon$ in the expression for T, since T appears twice in this sandwiching formula. This factor of $1/2$ also appears in the formula for rotation, where the factor $1/2$ multiplies the angle θ because both q and q^* appear in the sandwiching map for rotation.

Nevertheless, this sandwiching formula for translation is still not quite compatible with the sandwiching formula for rotation because a conjugate appears in the sandwiching formula (4.1) for rotation which is absent from this sandwiching formula (4.2) for translation. To overcome this incompatibility between Equations (4.1) and (4.2), we shall invoke the cross variant of the conjugate. Since $t = t_1 i + t_2 j + t_3 k$ is a pure quaternion, $t^* = -t$, so $(t\varepsilon)^\dagger = t\varepsilon$. Thus, for the dual quaternions representing translation

$$T^\dagger = O + \frac{d}{2}t\varepsilon = T.$$

Furthermore, for dual quaternions representing rotations, the coefficient of ε is zero, so

$$q^\dagger = q^*.$$

Therefore, we have the following compatible sandwiching formulas for translation and rotation:

Theorem 4.1 (Rotation and Translation)

Let N and t be unit vectors in the quaternion algebra. Then

1. *Rotation by the angle θ around the line through the origin O parallel to the unit normal vector N:*

$$P \mapsto qPq^* = qPq^\dagger \quad q = q(N, \theta/2) = \cos(\theta/2)O + \sin(\theta/2)N$$

2. *Translation by the distance d in the direction of the unit vector $t\varepsilon$:*

$$P \mapsto TPT = TPT^\dagger \quad T = T(t, d/2) = O + \frac{d}{2}t\varepsilon$$

Rigid motions can now be computed by composing rotations and translations using dual quaternion multiplication:

$$\text{Rotation followed by Translation} : P \mapsto T\left(qPq^{\dagger}\right)T^{\dagger} = \left(Tq\right)P\left(Tq\right)^{\dagger}$$

$$\text{Translation followed by Rotation} : P \mapsto q\left(TPT^{\dagger}\right)q^{\dagger} = \left(qT\right)P\left(qT\right)^{\dagger}.$$

Thus, the composite of a rotation and a translation can be represented by the product of the corresponding dual quaternions.

Exercises

1. Let $T = T(t,d/2)$ be a dual quaternion representing a translation. Show that $T^{-1} = \overline{T} = T^{*}$.
2. Show that $q(N,\theta/2)$ and $T(t,d/2)$ are unit dual quaternions.
3. Using the Taylor expansion of e^{x}, show that
   a. $q(N,\theta/2) = e^{\theta N/2}$
   b. $T(t,d/2) = e^{dt\varepsilon/2}$

1.4.2 Rotations about Arbitrary Lines and Screw Transformations

Rotations in the quaternion algebra are around lines through the origin. Adapting the quaternion approach for rotation to dual quaternions, we can rotate a point P by the angle θ around a line L passing through the origin O by sandwiching P between a unit quaternion and its conjugate (Theorem 4.1). Since in the space of dual quaternions we also have translations, we can rotate points P by the angle θ around a line L passing through an arbitrary point C by first translating the line L and the point P by the vector $O - C$, then rotating the translated point P' about the translated line L' that passes through the origin O and is parallel to the original line L, and finally translating the resulting point by the vector $C - O$ back to its true position. But this approach requires three transformations: two translations and one rotation. Surely by composing these three transformations using multiplication of dual quaternions, there is a single, simple dual quaternion that achieves the same result. Here's how.

Theorem 4.2 (Rotation about an Arbitrary Line in 3-Dimensions)

Let C be a dual quaternion representing a point, and let N be a dual quaternion representing a unit normal vector. Set

$$R = R(N, \theta, C) = \cos(\theta/2)O + \sin(\theta/2)(N + (C - O) \times N).$$

Then the map $P \mapsto RPR^+$ rotates the point P by the angle θ about the line L passing through the point C parallel to the unit normal vector N.

Proof: Let $t\varepsilon = C - O$, $d = |C - O|$, and consider the transformations represented by

$$T = T(t/d, d/2) = O + \frac{1}{2}t\varepsilon \qquad \{translation\}$$

$$q = q(N, \theta/2) = \cos(\theta/2)O + \sin(\theta/2)N \quad \{rotation\}$$

By the discussion in the paragraph preceding this theorem (translate the axis to pass through the origin, rotate around the axis through the origin, translate back), we only need to show that $TqT^{-1} = R$. But $T^{-1} = O - \frac{1}{2}t\varepsilon$, so

$$TqT^{-1} = \left(O + \frac{1}{2}t\varepsilon\right)q\left(O - \frac{1}{2}t\varepsilon\right)$$

$$= \left(q + \frac{1}{2}tq\varepsilon\right)\left(O - \frac{1}{2}t\varepsilon\right)$$

$$= q + \frac{1}{2}tq\varepsilon - \frac{1}{2}qt\varepsilon$$

$$= q + \frac{tq - qt}{2}\varepsilon \qquad\qquad \{q = \cos(\theta/2)O + \sin(\theta/2)N\}$$

$$= \cos(\theta/2)O + \sin(\theta/2)N + \sin(\theta/2)\frac{tN - Nt}{2}\varepsilon \quad \{\text{since } tO = Ot\}$$

$$= \cos(\theta/2)O + \sin(\theta/2)(N + (t \times N)\varepsilon)$$

$$\{\text{quaternion formula for cross product}\}$$

$$= \cos(\theta/2)O + \sin(\theta/2)(N + (C - O) \times N)$$

$$= R(N, \theta, C). \lozenge$$

Remark 4.3

Notice that in the expression for $R(N,\theta,C)$ in Theorem 4.2, the coefficient of $\sin(\theta/2)$ is $L = N + (C - O) \times N$. This coefficient is precisely the representation in terms of Plucker coordinates in the algebra of dual quaternions for the line L that serves as the axis of rotation.

We can generalize Theorem 4.1 still further to screw transformations about arbitrary lines L by composing a rotation around the line L with a translation in the direction parallel to the line L.

Theorem 4.4 (Screw Transformations— Simultaneous Rotation and Translation)

Let C be a dual quaternion representing a point, and let N be a dual quaternion representing a unit normal vector. Set

$$R = R(N,\theta,C) = \cos(\theta/2)O + \sin(\theta/2)\big(N + (C-O) \times N\big)$$

$$S = S(N,\theta,C,d) = R(N,\theta,C) + \frac{1}{2}d\big(\cos(\theta/2)N - \sin(\theta/2)O\big)\varepsilon.$$

Then the map $P \mapsto SPS^{\dagger}$ first rotates the point P by the angle θ about the line L passing through the point C parallel to the unit normal vector N and then translate the resulting point in the direction N by the distance d.

Proof: By Theorem 4.2 the dual quaternion $R = R(N,\theta,C)$ represents rotation by the angle θ about the line L passing through the point C parallel to the unit normal vector N. Furthermore, by Theorem 4.1 the dual quaternion $T = T(N,d/2)$ represents translation in the direction of the unit vector $N\varepsilon$ parallel to the unit normal vector N by the distance d. Therefore, we only need to show that $TR = S$. Let

$$q = q(N,\theta/2) = \cos(\theta/2)O + \sin(\theta/2)N$$

$$C = O + t\varepsilon$$

Then

$$R = \cos(\theta/2)O + \sin(\theta/2)\big(N + (C-O) \times N\big) = q + \sin(\theta/2)(t \times N)\varepsilon$$

$$TR = \left(O + \frac{1}{2}dN\varepsilon\right)\big(q + \sin(\theta/2)(t \times N)\varepsilon\big) = R + \frac{1}{2}dNq\varepsilon.$$

Moreover, since N is a unit vector in the quaternion algebra, $N^2 = -O$. Hence

$$Nq = N\left(\cos(\theta/2)O + \sin(\theta/2)N\right) = \cos(\theta/2)N - \sin(\theta/2)O.$$

$$TR = R + \frac{1}{2}d\left(\cos(\theta/2)N - \sin(\theta/2)O\right)\varepsilon = S. \lozenge$$

The screw transformations contain both the translations and the rotations as special cases. When $d = 0$ we have $S(N,\theta,C,0) = R(N,\theta,C)$, and when $\theta = 0$ we have $S(N,0,C,d) = T(N,d/2)$ (see Exercise 7). We shall see shortly (Corollary 4.7) that the screw transformations are the most general kinds of orientation preserving rigid motions. Reflections in arbitrary planes will be discussed in Section 1.4.6.

The transformation $S = S(N,\theta,C,d)$ rotates points by the angle θ about the line L passing through the point C parallel to the unit normal vector N and then translate the resulting point in the direction of the unit vector $N\varepsilon$ by the distance d. If we let $d = d(\theta)$ be a function of θ or more generally if we let $\theta = \theta(t)$ and $d = d(t)$ be functions of the time t and set $S(t) = S\left(N,\theta(t),C,d(t)\right)$, then the map $P \mapsto S(t)PS^*(t)$ represents a simultaneous rotation and translation, a screw motion around the line L; see Figure 4.1.

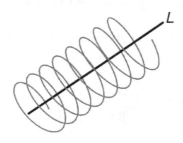

FIGURE 4.1
Screw motion around the line L.

Exercises

1. Consider the following two transformations:

$$q = q(N,\theta/2) = \cos(\theta/2)O + \sin(\theta/2)N \quad \{rotation\}$$

$$T = T(t,d/2) = O + \frac{d}{2}t\varepsilon \quad\quad\quad \{translation\}$$

  a. Show that:

$$Tq - qT = d\,\sin(\theta/2)(t \times N)\varepsilon.$$

  b. Conclude from part a that rotation and translation do not commute unless the translation direction t is parallel to the direction N of the axis of rotation.

2. Using the notation of Theorems 4.2 and 4.4, show that:

  a. $R(N,\theta,C)$ is independent of the choice of the point C on the line L.

  b. $S(N,\theta,C,d)$ is independent of the choice of the point C on the line L.

3. Using the notation of Theorems 4.2 and 4.4, show that $R(N,\theta,C)$ and $S(N,\theta,C,d)$ are unit dual quaternions. (Hint: Use Exercise 2 from Section 1.4.1.)

4. Let N and t be unit vectors in the quaternion algebra, and set $T = O + \dfrac{1}{2}t\varepsilon$. Show that

$$TNT^{-1} = N + (t \times N)\varepsilon.$$

5. Consider two points P_1, P_2 and two transformations:

$$q = q(N,\theta/2) = \cos(\theta/2)O + \sin(\theta/2)N \quad \{rotation\}$$

$$T_d = T(t,d/2) = O + \frac{d}{2}t\varepsilon \qquad\qquad \{translation\}$$

Show that:

  a. $q(P_1 P_2)q^* = \left(qP_1q^*\right)\left(qP_2q^*\right)$

  b. $T_d(P_1 P_2)T_d = \left(T_{d/2}P_1T_{d/2}^\dagger\right)\left(T_{d/2}P_2T_{d/2}^\dagger\right)$

Conclude that:

  c. Rotation commutes with dual quaternion multiplication for points.

  d. Translation does not commute with dual quaternion multiplication for points.

6. Let C be a dual quaternion representing a point, and let N be a dual quaternion representing a unit normal vector. Set

$$R = N + \big((C - O) \times N\big)\varepsilon.$$

Show that the map $P \mapsto RPR^\dagger$ rotates P by the angle π around the line L passing through the point C parallel to the unit normal vector N.

7. Let C be a dual quaternion representing a point, and let N be a dual quaternion representing a unit normal vector. Using the notation of Theorems 4.2 and 4.4, show that:

   a. $S(v\varepsilon)S^{\dagger} = R(v\varepsilon)R^{\dagger} = q(v\varepsilon)q^{*}$

   b. $S(N,0,C,d) = T(N,d/2)$

   c. $S(N,\theta,C,0) = R(N,\theta,C)$

8. Let r denote a rigid motion (i.e. a translation or rotation), and let s denote a screw transformation. Show that

   a. the conjugates r^{*}, $\overline{r}$, $r^{\dagger}$ are rigid motions.

   b. the conjugate s^{*} is a screw transformation, but the conjugate $\overline{s}$ is not screw a transformation.

9. Show that rotating a point by the angle θ about the line L passing through the point C parallel to the unit normal vector N is equivalent to first rotating the point by the angle θ around the line L' passing through the origin O parallel to the unit normal vector N, and then translating the point by the vector

$$v = \sin(\theta/2)\big(\cos(\theta/2)(C-O)\times N + \sin(\theta/2)(C-O)$$

$$-\sin(\theta/2)\big((C-O)\cdot N\big)N\big)$$

That is, using the notation of Theorem 4.2, show that

$$R(N,\theta,C) = T(t,d/2)q(N,\theta/2)$$

$$q(N,\theta/2) = \cos(\theta/2)O + \sin(\theta/2)N$$

$$\frac{1}{2}dt\varepsilon = \sin(\theta/2)\big(\cos(\theta/2)(C-O)\times N + \sin(\theta/2)(C-O)$$

$$-\sin(\theta/2)\big((C-O)\cdot N\big)N\big).$$

1.4.3 Screw Transformations and Rigid Motions

The goal of this section is to show that the screw transformations are the most general kinds of rigid motions, that all rigid motions—that is, all composites of rotations about arbitrary lines and translations in arbitrary directions—are screw transformations.

Theorem 4.5

Every rotation around a line through the origin followed by a translation in an arbitrary direction is a screw transformation.

Proof: Consider a rotation q by the angle θ around a line through the origin parallel to the unit normal vector N, and a translation T in an arbitrary direction by the vector $t\varepsilon$. Then

$$q = q(N,\theta/2) = \cos(\theta/2)O + \sin(\theta/2)N$$

$$T = O + \frac{1}{2}t\varepsilon.$$

We are going to show that

$$Tq = S(N,\theta,C,d).$$

where

$$C = O + \frac{1}{2}\left(\cot(\theta/2)N \times t + t\right)\varepsilon$$

$$d = t \cdot N.$$

Taking the product of T and q and invoking the identity $t \times N - (t \cdot N)O = tN$ (see Section 1.2.1, Exercise 1a) yields

$$Tq = \left(O + \frac{1}{2}t\varepsilon\right)\left(\cos(\theta/2)O + \sin(\theta/2)N\right)$$

$$= \cos(\theta/2)O + \sin(\theta/2)N + \frac{1}{2}\cos(\theta/2)t\varepsilon + \frac{1}{2}\sin(\theta/2)tN\varepsilon$$

$$= \cos(\theta/2)O + \sin(\theta/2)\left(N + \frac{1}{2}\cot(\theta/2)t\varepsilon - \frac{1}{2}(t \cdot N)O\varepsilon + \frac{1}{2}(t \times N)\varepsilon\right)$$

To simplify this expression, decompose t into components $t_{\parallel}$ and $t_{\perp}$, parallel and perpendicular to N, so that

$$t = t_{\perp} + t_{\parallel}$$

$$t_{\parallel} = (t_{\parallel} \cdot N)N = (t \cdot N)N$$

$$t_\| N = -(t \cdot N)O$$

$$t \times N = t_\perp \times N$$

Now

$$Tq = \cos(\theta/2)O + \sin(\theta/2)\left(N + \frac{1}{2}\cot(\theta/2)(t_\perp + t_\|)\varepsilon \right.$$

$$\left. -\frac{1}{2}(t \cdot N)O\varepsilon + \frac{1}{2}(t_\perp \times N)\varepsilon \right)$$

$$= \cos(\theta/2)O + \sin(\theta/2)\left(N + \frac{1}{2}\cot(\theta/2)t_\perp\varepsilon + \frac{1}{2}(t_\perp \times N)\varepsilon \right)$$

$$+\frac{1}{2}\cos(\theta/2)t_\|\varepsilon - \frac{1}{2}\sin(\theta/2)(t \cdot N)O\varepsilon$$

$$= \cos(\theta/2)O + \sin(\theta/2)\left(N + \frac{1}{2}\cot(\theta/2)t_\perp\varepsilon + \frac{1}{2}(t_\perp \times N)\varepsilon \right)$$

$$+\frac{1}{2}(t \cdot N)\left(\cos(\theta/2)N - \sin(\theta/2)O\right)\varepsilon$$

Let

$$C = O + \frac{1}{2}\left(\cot(\theta/2)N \times t + t\right)\varepsilon.$$

Then

$$(C - O) \times N = \frac{1}{2}\left(\cot(\theta/2)(N \times t) \times N + t \times N\right)\varepsilon$$

$$= \frac{1}{2}\left(\cot(\theta/2)(N \times t_\perp) \times N + t_\perp \times N\right)\varepsilon$$

Moreover since $N, t_\perp, N \times t_\perp$ are mutually orthogonal and N is a unit vector

$$(N \times t_\perp) \times N = t_\perp$$

$$(C - O) \times N = \frac{1}{2}\left(\cot(\theta/2)t_\perp + t_\perp \times N\right)\varepsilon.$$

Therefore by Theorem 4.4

$$Tq = \cos(\theta/2)O + \sin(\theta/2)\big(N + (C-O)\times N\big)$$

$$+ \frac{1}{2}(t \cdot N)\big(\cos(\theta/2)N - \sin(\theta/2)O\big)\varepsilon$$

$$= R(N,\theta,C) + \frac{1}{2}(t \cdot N)\big(\cos(\theta/2)N - \sin(\theta/2)O\big)\varepsilon$$

$$= S(N,\theta,C,t \cdot N). \Diamond$$

Lemma 4.6

Every rigid motion is equivalent to a single rotation about the origin followed by a single translation.

Proof: By definition a rigid motion is the product (composite) of rotations about arbitrary lines and translations in arbitrary directions. Moreover, by the proof of Theorem 4.2, every rotation about an arbitrary line is the product of a rotation about a line through the origin sandwiched between two translations. Therefore, to establish this result, we shall first prove the following three straightforward assertions:

1. The product of two rotations about the origin is another rotation about the origin.
2. The product of two translations is another translation.
3. The product of a translation followed by a rotation about the origin is equivalent to first performing the rotation about the origin and then performing a (different) translation.
   1. Recall that a quaternion q represents a rotation about the origin if and only if q is a unit quaternion. But the product of two unit quaternions is another unit quaternion. Therefore, the product of two rotations about the origin is another rotation about the origin.
   2. Consider two translations

$$T_1 = O + t_1\varepsilon$$

$$T_2 = O + t_2\varepsilon.$$

        Then

$$T_1T_2 = (O + t_1\varepsilon)(O + t_2\varepsilon) = O + (t_1 + t_2)\varepsilon$$

is another translation. Therefore, the product of two translations is another translation.

3. Consider a translation $T = O + t\varepsilon$ followed by a rotation q about the origin. Then

$$qT = q(O + t\varepsilon) = (q + qt\varepsilon) = (O + qtq^*\varepsilon)q.$$

Now qtq^* is just the vector t rotated by q. Therefore, the product of a translation followed by a rotation about the origin is equivalent to first performing the rotation about the origin and then performing a (different) translation.

If follows from the third assertion that for any product of rotations about the origin and translations in any order, we can first perform a product of rotations about the origin and then perform a product of translations. Since the product of rotations about the origin is a single rotation about the origin and the product of translations is another translation, every rigid motion is equivalent to one rotation about the origin followed by one translation. ◊

Corollary 4.7 (Chasles' Theorem)

Every rigid motion is equivalent to a screw transformation.

Proof: This result follows immediately from Theorem 4.5 and Lemma 4.6. ◊

Proposition 4.8

Every screw transformation is a rigid motion.

Proof: Every screw transformation is the product of a rotation about an arbitrary axis line and a translation along the axis of rotation. Therefore, every screw transformation is the product of two rigid motions and hence a rigid motion. ◊

Corollary 4.9

The product of two screw transformations is a screw transformation.

This result follows immediately from Proposition 4.8 and Corollary 4.7 since the product of two rigid motions is another rigid motion. ◊

Exercise

1. By Lemma 4.6 every rigid motion is equivalent to a rotation around a line through the origin followed by a single translation. Let

   q = quaternion that rotates points by the angle θ around the line through the origin parallel to the unit normal vector N.

   T = dual quaternion that translates points on the line through the origin parallel to the unit normal vector N to another line L parallel to the unit normal vector N.

   R = dual quaternion that rotates points by the angle θ around the line L.

   a. Show that $R \neq Tq$.

   b. Explain why the equality in part a fails.

   c. What translations T' needs to be inserted between T and q so that $R = TT'q$.

   d. Explain why the equality in part c does not violate Lemma 4.6.

1.4.4 Rotation and Translation on Planes

We have shown how to use sandwiching to compute rotations and translations on points and vectors. But what about planes? How does sandwiching with the rotations $q(N,\theta/2)$ or the translations $T(N,d/2)$ affect planes π?

Rotations first. Consider a plane $\pi = v + dO\varepsilon$ and a rotation $q = q(N,\theta/2) = \cos(\theta/2)O + \sin(\theta/2)N$. Then sandwiching between q and q^* yields the maps

$$v \mapsto qvq^*$$

$$dO\varepsilon \mapsto q(dO\varepsilon)q^* = dO\varepsilon.$$

Therefore, the sandwiching map

$$\pi \mapsto q\pi q^* = qvq^* + dO\varepsilon$$

rotates the normal vector v of the plane π by the angle θ around the axis line through the origin O parallel to the unit normal vector N and leaves the signed distance $d/|v|$ to the origin unchanged.

Translation is a bit more subtle. Consider a translation $T = T(t, \delta/2)$ $= O + \dfrac{\delta}{2} t\varepsilon$. Then

$$T(dO\varepsilon)T = dO\varepsilon$$

$$T(v)T = \left(O + \frac{\delta}{2} t\varepsilon \right) v \left(O + \frac{\delta}{2} t\varepsilon \right) = \left(v + \frac{\delta}{2} tv\varepsilon \right) \left(O + \frac{\delta}{2} t\varepsilon \right)$$

$$= v + \delta \frac{tv + vt}{2} \varepsilon = v - \delta(t \cdot v) O\varepsilon.$$

Thus, the map

$$\pi \mapsto T\pi T = T(v + dO\varepsilon)T = v + (d - \delta(t \cdot v))O\varepsilon = \pi - \delta(t \cdot v)O\varepsilon$$

leaves the normal vector v to the plane π unchanged, but changes the signed distance of the plane π to the origin by $-\dfrac{\delta(t \cdot v)}{|v|}$—that is, translates the plane π along the normal direction ν by $-\dfrac{\delta(t \cdot v)}{|v|}$.

Notice two things about translation. First, translation behaves differently on normal vectors than on ordinary vectors: ordinary vectors $u\varepsilon$ are unaffected by translation, but translation introduces an $O\varepsilon$ term to normal vectors v. This term leaves the normal vector itself unchanged, but changes the distance of the plane from the origin. Second, only the component $t \cdot v$ of the translation vector $t\varepsilon$ along the normal direction ν matters. This property makes sense because translating a plane in a direction perpendicular to the normal to the plane maps the plane onto itself. Thus, the map $\pi \mapsto T\pi T$ actually translates the plane π by the vector $t\varepsilon$, but the equation of the plane—and therefore the representation of the plane in the space of dual quaternions—is affected only along the component of the translation vector in the normal direction.

To summarize, we have the following results for rotation and translation on planes, similar to our results for rotation and translation on points. Analogous results for planes also hold for rotations about arbitrary lines (see Exercise 3) and for screw transformations (see Exercise 4).

Theorem 4.10: (Rotation and Translation on Planes)

Let N and t be unit vectors in the quaternion algebra. Then

1. *Rotation by the angle θ around the line through the origin O parallel to the unit normal vector N:*

$$\pi \mapsto q\pi q^* = q\pi q^\dagger \quad q = q(N, \theta/2) = \cos(\theta/2)O + \sin(\theta/2)N.$$

2. *Translation by the distance d in the direction $t\varepsilon$:*

$$\pi \mapsto T\pi T = T\pi T^\dagger \quad T = T(t, d/2) = O + \frac{d}{2}t\varepsilon.$$

Now we can apply translations and rotations independently to points and to planes. But are these transformations compatible? Do these transformations always map a point P on a plane π to a point P' on the transformed plane π'? Yes! To prove this result, we begin with a simple lemma relating the dot product and sandwiching.

Lemma 4.11

Let p, r be arbitrary dual quaternions and let q, t be unit dual quaternions. Then

1. $\left(qpq^* \right) \cdot \left(qrq^* \right) = p \cdot r$
2. $\left(tpt \right) \cdot \left(trt \right) = p \cdot r$

Proof: We shall prove only the first equality; the second equality follows by a similar argument (see Exercise 1). To establish the first equality, recall that by the definition of the dot product

$$\left(qpq^* \right) \cdot \left(qrq^* \right) = \frac{\left(qpq^* \right)\left(qrq^* \right)^* + \left(qrq^* \right)\left(qpq^* \right)^*}{2}$$

Now we compute each term on the right-hand side:

a. $\left(qpq^* \right)\left(qrq^* \right)^* = \left(qpq^* \right)\left(qr^*q^* \right) = q\left(pr^* \right)q^*$

b. $\left(qrq^* \right)\left(qpq^* \right)^* = \left(qrq^* \right)\left(qp^*q^* \right) = q\left(rp^* \right)q^*$

Adding (a) and (b) and dividing by 2 yields

$$\left(qpq^* \right) \cdot \left(qrq^* \right) = q\left(\frac{pr^* + rp^*}{2} \right)q^* = (p \cdot r),$$

where the last equality follows because $p \cdot r$ is a dual real number and dual real numbers commute with dual quaternions. $\lozenge$

Theorem 4.12

Consider a point P and a plane π in the space of dual quaternions. Let

$$q = q(N, \theta/2) = \cos(\theta/2)O + \sin(\theta/2)N \quad \{rotation\}$$

$$T = T(t, d/2) = O + \frac{d}{2}t\varepsilon \qquad \{translation\}$$

Then

1. *the point P lies on the plane π $\Leftrightarrow$ the rotated point qPq^+ lies on the rotated plane $q\pi q^+$.*
2. *the point P lies on the plane π $\Leftrightarrow$ the translated point TPT^+ lies on the translated plane $T\pi T^+$.*

Proof: We shall prove only the first result; the second result follows by a similar argument (see Exercise 2). To derive the first result, recall that by Proposition 3.1:

$$\text{the point } P \text{ lies on the plane } \pi \Leftrightarrow P \cdot \pi = 0.$$

But by Lemma 4.11, since $q = q(N, \theta/2)$ is a unit dual quaternion (see Section 1.4.1, Exercise 2)

$$\left(qPq^*\right) \cdot \left(q\pi q^*\right) = P \cdot \pi$$

Therefore

$$\left(qPq^+\right) \cdot \left(q\pi q^+\right) = \left(qPq^*\right) \cdot \left(q\pi q^*\right) = 0 \Leftrightarrow P \cdot \pi = 0$$

Thus, the point P lies on the plane π $\Leftrightarrow$ the rotated point qPq^+ lies on the rotated plane $q\pi q^+$. $\Diamond$

A word of caution about the meaning of Theorem 4.12. Although we have proved that rigid motions always map a point P on a plane π to a point P' on the transformed plane π', nevertheless it may happen that a rigid motion leaves the plane unchanged but moves the points on the plane. Consider the xy-plane. Rotating the xy-plane around the z-axis does not alter this plane (or the normal to this plane), but the points on the xy-plane are rotated around the z-axis to new points on the xy-plane. Thus, the points on the xy-plane are affected by this rotation, even though the plane itself is unchanged. Similarly, translating points in the x-direction does not alter xy-plane, but the points on the xy-plane are translated to new points on the xy-plane. Thus, once again the points on the xy-plane are affected by this translation, even though the plane itself is unchanged.

Exercises

1. Let p, r be arbitrary dual quaternions, and let t be a unit dual quaternion. Prove that

$$(tpt) \cdot (trt) = p \cdot r.$$

2. Consider a point P and a plane π in the space of dual quaternions. Let

$$T = T(t, \delta/2) = O + \frac{\delta}{2} t\varepsilon. \qquad \{translation\}$$

Verify that the point P lies on the plane $\pi \Leftrightarrow$ the translated point $TPT^{\dagger}$ lies on the translated plane $T\pi T^{\dagger}$.

3. State and prove an analogue of Theorem 4.2 (rotation about an arbitrary line) for planes.

4. State and prove an analogue of Theorem 4.4 (screw transformations) for planes.

1.4.5 Rotation and Translation on Lines

Recall that a line in 3-dimensions can be represented in at least five different ways:

1. Two points on the line.
2. One point on the line and a vector parallel to the line.
3. Two planes that intersect in the line.
4. Plucker coordinates.
5. Dual Plucker coordinates.

We already know how to rotate and translate points, vectors, and planes. Therefore, to rotate or translate a line l represented by points, vectors, or planes (methods 1,2,3), we simply rotate or translate the representative points, vectors, or planes to get a representation of the transformed line l' in terms of points, vectors, or planes.

Plucker coordinates and dual Plucker coordinates are just a bit more complicated. For lines represented using these coordinates, we can translate lines in arbitrary directions and rotate lines about lines through the origin, using the following theorems. Analogous results for lines also hold for rotations about arbitrary lines (see Exercise 3) and for screw transformations (see Exercise 4).

Theorem 4.13 (Rotation and Translation of Plucker Coordinates)

Let N and t be unit vectors in the quaternion algebra, and let $l = v + u\varepsilon = v + (P - O) \times v$ be the Plucker coordinates of the line l. Then

1. *Rotation by the angle θ around the line through the origin O parallel to the unit normal vector N:*

$$l \mapsto qlq^{-1} \quad q = q(N, \theta/2) = \cos(\theta/2)O + \sin(\theta/2)N.$$

2. *Translation by the distance d in the direction $t\varepsilon$:*

$$l \mapsto TlT^{-1} \quad T = T(t, d/2) = O + \frac{d}{2}t\varepsilon.$$

Proof:

1. By Theorems 4.1 and 4.10 the map $x \mapsto qxq^* = qxq^{-1}$ rotates points and planes by the angle θ around the line through the origin O parallel to the unit normal vector N. Therefore the maps

$$v \mapsto qvq^{-1}$$

$$u\varepsilon \mapsto qu\varepsilon q^{-1} = q((P - O) \times v)q^{-1} = q(P - O)q^{-1} \times qvq^{-1}$$

$$= (qPq^{-1} - O) \times qvq^{-1}$$

rotate the Plucker coordinates of the line l by the angle θ around the line through the origin O parallel to the unit normal vector N. Hence the map $l \mapsto qlq^{-1}$ rotates the line l by the angle θ around the line through the origin O parallel to the unit normal vector N.

2. Consider the maps

$$TvT^{-1} = \left(O + \frac{d}{2}t\varepsilon\right)v\left(O - \frac{d}{2}t\varepsilon\right) = v + \frac{d}{2}tv\varepsilon - \frac{d}{2}vt\varepsilon$$

$$= v + d\frac{tv - vt}{2}\varepsilon = v + d(t \times v)\varepsilon$$

$$Tu\varepsilon T^{-1} = u\varepsilon = (P - O) \times v$$

Adding these two equations, we find that

$$TlT^{-1} = TvT^{-1} + Tu\varepsilon T^{-1} = v + \left(d(t \times v)\varepsilon + (P - O) \times v\right)$$

$$= v + \left((P + dt\varepsilon - O) \times v\right)$$

Therefore, the map $l \mapsto TlT^{-1}$ leaves the normal vector v parallel to the line unchanged and translates the point P on the line by the distance d in the direction $t\varepsilon$. Hence the map $l \mapsto TlT^{-1}$ translates the line l by the distance d in the direction $t\varepsilon$. ◊

Theorem 4.14 (Rotation and Translation of Dual Plücker Coordinates)

Let N and t be unit vectors in the quaternion algebra, and let $l = v + u\varepsilon = v_1 \times v_2 + (d_1 v_2 - d_2 v_1)\varepsilon$ be the dual Plücker coordinates of the line l. Then

1. Rotation by the angle θ around the line through the origin O parallel to the unit normal vector N:

$$l \mapsto qlq^{-1} \quad q = q(N, \theta/2) = \cos(\theta/2)O + \sin(\theta/2)N.$$

2. Translation by the distance δ in the direction $t\varepsilon$:

$$l \mapsto TlT^{-1} \quad T = T(t, \delta/2) = O + \frac{\delta}{2}t\varepsilon$$

Proof:

1. By Theorems 4.1 and 4.10 the map $x \mapsto qxq^* = qxq^{-1}$ rotates points and planes by the angle θ around the line through the origin O parallel to the unit normal vector N. Therefore the maps

$$v \mapsto q(v_1 \times v_2)q^{-1} = (qv_1q^{-1}) \times (qv_2q^{-1})$$

$$u\varepsilon \mapsto qu\varepsilon q^{-1} = q(d_1 v_2 - d_2 v_1)\varepsilon q^{-1} = \left(d_1\left(q v_2 q^{-1}\right) - d_2\left(q v_1 q^{-1}\right)\right)\varepsilon$$

rotate the dual Plücker coordinates of the line l by the angle θ around the line through the origin O parallel to the unit normal vector N. Hence the map $l \mapsto qlq^{-1}$ rotates the line l by the angle θ around the line through the origin O parallel to the unit normal vector N.

2. Consider the maps

$$T v T^{-1} = \left(O + \frac{\delta}{2} t\varepsilon \right) v \left(O - \frac{\delta}{2} t\varepsilon \right) = v + \frac{\delta}{2} t v\varepsilon - \frac{\delta}{2} v t\varepsilon$$

$$= v + \delta \frac{tv - vt}{2}\varepsilon = v + \delta(t \times v)\varepsilon$$

$$T u\varepsilon T^{-1} = u\varepsilon = (d_1 v_2 - d_2 v_1)\varepsilon$$

and recall that (see Appendix)

$$t \times v = t \times (v_1 \times v_2) = (t \cdot v_2) v_1 - (t \cdot v_1) v_2.$$

Thus, adding the first two equations, we find that

$$T l T^{-1} = T v T^{-1} + T u\varepsilon T^{-1} = v + (d_1 - \delta(t \cdot v_1)) v_2 - (d_2 - \delta(t \cdot v_2)) v_1.$$

Therefore the map $l \mapsto T l T^{-1}$ leaves the normal vector v parallel to the line unchanged and translates the planes with normal vectors v_i by the distance $\dfrac{\delta(t \cdot v_i)}{|v_i|^2}$ in the direction v_i, $i = 1, 2$ —that is, translates the planes with normal vectors v_i by the distance δ in the direction $t\varepsilon$ (see Section 1.4.4). Hence the map $l \mapsto T l T^{-1}$ translates the line l by the distance δ in the direction $t\varepsilon$. $\lozenge$

Corollary 4.15 (Rotation and Translation on Lines)

Let N and t be unit vectors in the quaternion algebra, and let $l = v + u\varepsilon$ with $v \cdot u = 0$ represent the line l. Then

1. *Rotation by the angle θ around the line through the origin O parallel to the unit normal vector N:*

$$l \mapsto qlq^{-1} \qquad q = q(N, \theta/2) = \cos(\theta/2)O + \sin(\theta/2)N.$$

2. *Translation by the distance d in the direction tε:*

$$l \mapsto TlT^{-1} \quad T = T(t, d/2) = O + \frac{d}{2}t\varepsilon.$$

Proof: This result follows immediately from either Theorem 4.13 or Theorem 4.14, since l can always be expressed in terms of either Plucker coordinates or dual Plucker coordinates. ◊

Notice that the translation formula for lines represented in terms of either Plucker coordinates or dual Plucker coordinates is slightly different from the translation formula for points and planes: $T^{\dagger}$ is replaced by T^{-1}. Nevertheless, rigid motions on lines with these representations can still be computed by composing rotations and translations using dual quaternion multiplication.

Rotation followed by Translation : $l \mapsto T(qlq^{-1})T^{-1} = (Tq)l(Tq)^{-1}.$

Translation followed by Rotation : $l \mapsto q(TlT^{-1})q^{-1} = (qT)l(qT)^{-1}.$

Exercises

1. In the proofs of Theorems 4.13 and 4.14 we used the identity

$$q(v \times w)q^{-1} = (qvq^{-1}) \times (qwq^{-1}).$$

   Prove this identity (see too Section 1.2.1, Exercise 2a).

2. Consider a line l represented by Plucker or dual Plucker coordinates. Let

$$T = T(t, \delta/2) = O + \frac{\delta}{2}t\varepsilon. \quad \{translation\}$$

   Verify that a point P lies on the line $l \Leftrightarrow$ the translated point $TPT^{\dagger}$ lies on the translated line TlT^{-1}.

3. State and prove an analogue of Theorem 4.2 (rotation about an arbitrary line) for lines represented in terms of Plucker or dual Plucker coordinates.

4. State and prove an analogue of Theorem 4.4 (screw transformations) for lines represented in terms of Plucker or dual Plucker coordinates.

1.4.6 Reflections

Reflections are also rigid motions, albeit orientation reversing. It turns out that we can also use dual quaternions to compute reflections, but with a new twist. Let $\pi = v + dO\varepsilon$ be a plane with unit normal vector v. We are going to show that the map $x \mapsto -\pi x^* \pi$ reflects vectors, points, and planes in the plane π. We begin with the following lemma.

Lemma 4.16

Let π be a plane with unit normal vector v. Then the map $u \mapsto vuv$ reflects the (normal) vector u in the plane π.

Proof: Since v is a unit vector, it follows by the triple product formula for quaternion multiplication (see Section 1.2.1) that for any (normal) vector u,

$$vuv = u - 2(v \cdot u)v. \tag{4.3}$$

Now let $u_{\parallel}$ be the component of u parallel to v and let $u_{\perp}$ be the component of u perpendicular to v, so that $u = u_{\perp} + u_{\parallel}$. Then by Equation (4.3)

$$vu_{\perp}v = u_{\perp}$$

$$vu_{\parallel}v = -u_{\parallel} \tag{4.4}$$

$$vuv = u_{\perp} - u_{\parallel}$$

Thus, the component of u perpendicular to the unit normal vector v is unchanged, and the component of u parallel to the unit normal vector v is reflected in the plane π. Therefore, the map $u \mapsto vuv$ reflects the (normal) vector u in the plane π. $\lozenge$

Proposition 4.17

Let π be a plane with unit normal vector v. Then the map $u\varepsilon \mapsto -\pi(u\varepsilon)^* \pi$ reflects vectors $u\varepsilon$ in the plane π.

Proof: Let $\pi = dO\varepsilon + v$, where v is a unit normal vector. Then

$$-\pi(u\varepsilon)^* \pi = -v(u\varepsilon)^* v = (vuv)\varepsilon.$$

Therefore, this result follows immediately from Lemma 4.16. $\lozenge$

Proposition 4.18

Let π be a plane with unit normal vector v. Then the map $P \mapsto -\pi P^* \pi$ reflects points P in the plane π.

Proof: Let $\pi = v + dO\varepsilon$ and consider a point $C = O + u\varepsilon$ on the plane π. Then by Proposition 3.1

$$0 = C \cdot \pi = (u \cdot v + d)\varepsilon.$$

Therefore, since v is a unit normal vector,

$$\pi C^* \pi = (v + dO\varepsilon)(O - u\varepsilon)(v + dO\varepsilon) = vOv - v(u\varepsilon)v + 2dvO\varepsilon$$

$$= -O + \big(2(u \cdot v)v - u\big)\varepsilon + 2dv\varepsilon = -O - u\varepsilon + 2(u \cdot v + d)v\varepsilon$$

$$= -O - u\varepsilon = -C$$

Thus $-\pi C^* \pi \equiv C$, so the map $C \mapsto -\pi C^* \pi$ maps points on the plane π into themselves. Now consider an arbitrary point P. Then there is a point C on the plane π and a real number s such that $P = C + sv\varepsilon$. Hence, since $v^2 = -1$,

$$-\pi P^* \pi = -\pi(C^* + sv^* \varepsilon)\pi = -\pi C^* \pi + \pi(sv\varepsilon)\pi = C + v(sv\varepsilon)v = C - sv\varepsilon.$$

Thus, the map $P \mapsto -\pi P^* \pi$ reflects points P in the plane π. ◊

Proposition 4.19

Let π be a plane with unit normal vector v. Then the map $\rho \mapsto -\pi\rho^* \pi$ reflects planes ρ in the plane π.

Proof: Let ρ be a plane with normal vector μ, and let C be a point on the intersection of the planes π and ρ. Then by Corollary 3.2

$$\pi = v - C \cdot v$$

$$\rho = \mu - C \cdot \mu$$

Therefore since $C \cdot v$ and $C \cdot \mu$ are scalars times ε (see the proof of Corollary 3.2)

$$\pi\rho^* \pi = (v - C \cdot v)(\mu^* - C \cdot \mu)(v - C \cdot v)$$

$$= v\mu^* v - v(C \cdot \mu)v - (C \cdot v)(v\mu^* + \mu^* v)$$

Now let's consider each term on the right-hand side.

a. $v\mu^* v = -vuv = 2(v \cdot \mu)v - \mu$

b. $-v(C \cdot \mu)v = C \cdot \mu$ (because $C \cdot \mu$ is a scalar times ε)

c. $-(C \cdot v)(v\mu^* + \mu^* v) = 2(C \cdot v)\left(\dfrac{vu + uv}{2}\right) = -2(C \cdot v)(v \cdot \mu)$

Adding (a), (b), (c), we find that

$$\pi\rho^*\pi = \left(2(v \cdot \mu)v - \mu\right) - C \cdot \left(2(v \cdot \mu)v - \mu\right)$$

$$-\pi\rho^*\pi = \left(\mu - 2(v \cdot \mu)v\right) - C \cdot \left(\mu - 2(v \cdot \mu)v\right).$$

Thus $-\pi\rho^*\pi$ is the plane through the point C with normal vector $\mu - 2(v \cdot \mu)v$. But by Equation (4.3),

$$\mu - 2(v \cdot \mu)v = vuv$$

and by Lemma 4.16, the map $u \mapsto vuv$ reflects the normal vector u in the plane π. Therefore, the map $\rho \mapsto -\pi\rho^*\pi$ reflects planes ρ in the plane π. ◊

Theorem 4.20

Let π be a plane with unit normal vector v. Then the map $x \mapsto -\pi x^*\pi$ reflects vectors, points, and planes in the plane π.

Proof: This result follows immediately from Propositions 4.17, 4.18, and 4.19. ◊

The formula for reflecting lines is somewhat different from the formula in Theorem 4.20 for reflecting vectors, points, and planes. Here is the formula for reflecting lines in planes.

Theorem 4.21

Let $\pi = dO\varepsilon + w$ be a plane with unit normal vector w, and let $l = v + u\varepsilon$ with $v \cdot u = 0$ represent a line l. Then the map

$$l \mapsto \pi l^*\pi^{-1} = \pi v^*\pi^{-1} - \pi u^*\varepsilon\pi^{-1}$$

reflects lines l in the plane π.

Proof: We need to compute $\pi v^* \pi^{-1}$ and $\pi u^* \varepsilon \pi^{-1}$. For the first term observe that

$$\pi v^* \pi^{-1} = (dO\varepsilon + w)v^*(dO\varepsilon - w) = \left(dv^*\varepsilon + wv^*\right)(dO\varepsilon - w)$$

$$= d\left(wv^* - v^*w\right)\varepsilon - wv^*w = -2d(w \times v)\varepsilon + wvw.$$

Since w is a unit normal, it follows from Lemma 4.16 that the term wvw reflects the (normal) vector v in the plane π. We still have an extraneous term $-2d(w \times v)\varepsilon$, but this term will eventually cancel out when we compute $\pi u^* \varepsilon \pi^{-1}$.

To begin the computation of $\pi u^* \varepsilon \pi^{-1}$, observe that since $\varepsilon^2 = 0$ and $-w = w^{-1}$,

$$\pi\left(u^*\varepsilon\right)\pi^{-1} = (dO\varepsilon + w)\left(u^*\varepsilon\right)(dO\varepsilon - w) = w\left(u^*\varepsilon\right)w^{-1}.$$

To compute $w(u^*\varepsilon)w^{-1}$, recall that we can always write the line l in terms of Plucker coordinates so that $u\varepsilon = (P - O) \times v$, where P is a point on the line l. Hence

$$u^*\varepsilon = \left((P - O) \times v\right)^*.$$

Now recall from Section 1.2.1, Exercises 2 and 3 that for any vectors a, b

$$(a \times b)^* = -a^* \times b^*$$

$$w(a \times b)^* w^{-1} = -w\left(a^* \times b^*\right)w^{-1} = -\left(wa^*w^{-1}\right) \times \left(wb^*w^{-1}\right).$$

Letting $a = P - O$ and $b = v$ yields

$$w\left(u^*\varepsilon\right)w^{-1} = -\left(w(P - O)^* w^{-1}\right) \times \left(wv^*w^{-1}\right).$$

To simplify the factors on the right-hand side, let $P = O + v_P\varepsilon$. Then

$$wv^*w^{-1} = wvw$$

$$-w(P - O)^* w^{-1} = w\left(P^* - O\right)w = -(wv_Pw)\varepsilon.$$

Hence

$$w\left(u^*\varepsilon\right)w^{-1} = -\left(w(P - O)^* w^{-1}\right) \times \left(wv^*w^{-1}\right) = -(wv_Pw)\varepsilon \times (wvw)$$

$$-\pi(u^*\varepsilon)\pi^{-1} = -w\left(u^*\varepsilon\right)w^{-1} = (wv_Pw)\varepsilon \times (wvw).$$

But

$$\pi P^* \pi = (dO\varepsilon + w)(O - v_P\varepsilon)(dO\varepsilon + w) = (dO\varepsilon + w - wv_P\varepsilon)(dO\varepsilon + w)$$

$$= 2dw\varepsilon - wv_Pw\varepsilon - O.$$

Solving for $wv_Pw\varepsilon$, we find that

$$wv_Pw\varepsilon = -\pi P^* \pi - O + 2dw\varepsilon.$$

Thus

$$-\pi(u^*\varepsilon)\pi^{-1} = \left(-\pi P^* \pi - O + 2dw\varepsilon\right) \times (wvw)$$

$$= \left(-\pi P^* \pi - O\right) \times (wvw) + (2dw\varepsilon) \times (wvw).$$

Moreover, by Equation (4.3)

$$(2dw\varepsilon) \times (wvw) = (2dw\varepsilon) \times \left(v - 2(w \cdot v)w\right) = 2d(w \times v)\varepsilon.$$

Therefore, finally we arrive at

$$\pi v^* \pi^{-1} - \pi\left(u^*\varepsilon\right)\pi^{-1} = vwv - 2d(w \times v)\varepsilon + \left(-\pi P^* \pi - O\right) \times (wvw) + 2d(w \times v)\varepsilon$$

$$= vwv + \left(-\pi P^* \pi - O\right) \times (vwv).$$

By Lemma 4.16 the expression wvw is the reflection of the (normal) vector v in the plane π, and by Proposition 4.18 the expression $-\pi P^* \pi$ is the reflection of the point P in the plane π. Therefore, the map $l \mapsto \pi l^\dagger \pi^{-1} = \pi v^* \pi^{-1} - \pi u^* \varepsilon \pi^{-1}$ reflects the Plücker coordinates of the line l, and hence the line l, in the plane π. ◊

Exercises

1. Let P be a point, let ρ be a plane, and let π be a plane with a unit normal vector.
   a. Show that
      i. $P^* \cdot \rho^* = P \cdot \rho$
      ii. $\left(\pi P^* \pi\right) \cdot \left(\pi \rho^* \pi\right) = P \cdot \rho$
   b. Conclude that $\left(\pi P^* \pi\right) \cdot \left(\pi \rho^* \pi\right) = 0 \Leftrightarrow P \cdot \rho = 0$. Interpret this result geometrically.

2. In the proof of Proposition 4.19, we assume that the planes ρ and π intersect. Show that Proposition 4.19 remains valid when the planes ρ and π are parallel. That is, show that the map $\rho \mapsto -\pi\rho^*\pi$ reflects the plane ρ in the plane π, even when the plane ρ is parallel to the plane π.

3. What is the effect of the map $x \mapsto x^*$ on

   a. vectors, points, planes, and lines

   b. translations, rotations, and reflections.

4. Explain how to reflect a line l in a plane π with unit normal vector w, when the line l is represented in terms of:

   a. Two points on the line.

   b. One point on the line and a vector parallel to the line.

   c. Two planes that intersect in the line.

5. Prove Theorem 4.21 using dual Plucker coordinates in place of Plucker coordinates.
   Which proof do you think is simpler: using Plucker coordinates or dual Plucker coordinates?

1.5

Rigid Motions as Rotations in 8-Dimensions

One of the main insights regarding quaternions is that rotations, reflections, and perspective projections in 3-dimensions can all be modeled by rotations in 4-dimensions [8, Chapter 6]. Can we extend this insight to dual quaternions? Are translations, rotations, and more generally all rigid motions, rotations in 8-dimensions? Can perspective projections also be modeled by rotations in 8-dimensions?

Rotations in n-dimensions are linear isometries, linear maps that preserve angles and lengths. So to determine if rigid motions in 3-dimensions are really represented by rotations in 8-dimensions, we need to consider how linear isometries can be represented by dual quaternions.

1.5.1 Rigid Motions as Linear Isometries in 8-Dimensions

One of the key formulas explaining why unit quaternions represent rotations in 4-dimensions is that the length of the product of two quaternions is equal to the product of their lengths. The same identity holds for two dual quaternions p, q in 8-dimensions, since by Equation (2.5)

$$|pq| = |p||q|.$$

Thus, multiplication by a unit dual quaternion q on the right is a linear isometry in 8-dimensions:

$$p \to pq \text{ is linear } - (p_1 + p_2)q = p_1q + p_2q.$$

$$p \to pq \text{ is an isometry } - |pq| = |p||q| = |p|.$$

Similarly, multiplication by a unit dual quaternion q on the left is also a linear isometry in 8-dimensions. But linear isometries are rotations. Indeed,

DOI: 10.1201/9781003398141-6

we have seen in Section 2.3 that multiplication by unit dual quaternions preserves both lengths and angles. Since the dual quaternions representing translations and rotations are unit dual quaternions (see Section 1.4.1, Exercise 2), it follows that translations, rotations, and therefore all rigid motions, in particular screw transformations in 3-dimensions, are rotations in 8-dimensions.

Unit norm dual quaternions $q = q_1 + q_2\varepsilon$ must satisfy two conditions: $|q_1|^2 = O$ and $q_1 \cdot q_2 = 0$ [see Equation (2.4)]. Therefore, unit norm dual quaternions form a 6-dimensional subspace of the 8-dimensional space of dual quaternions. Moreover,

$$\textit{Rotations: } q = q(N, \theta/2) = \cos(\theta/2)O + \sin(\theta/2)N$$

$$\textit{Translation: } T = T(t, d/2) = O + \frac{d}{2}t\varepsilon,$$

so the rotations form a 3-dimensional subspace (3 angles) and the translations form another 3-dimensional subspace (3 directions) of the dual quaternions. Thus, altogether the rigid motions form a 6-dimensional subspace of the 8-dimensional space of dual quaternions.

Exercise

1. Show that the dual quaternions representing reflections also have unit norm.

1.5.2 Renormalization

If we compose several rigid motions represented by unit dual quaternions using dual quaternion multiplication with floating point arithmetic, then due to floating point errors eventually the products will no longer be unit dual quaternions. Objects will be distorted if we use these unnormalized dual quaternions to transform a scene. Therefore, we need a method to renormalize dual quaternions.

Quaternions can be renormalized by dividing by their length. Dual quaternions can also be renormalized by dividing by their length. But recall that whereas the length of a quaternion is a real number, the length of a dual quaternion is a dual real number. Therefore, the dual quaternion $p = q / |q|$ is not in the immediately useable form $p = p_1 + p_2\varepsilon$.

Here we shall derive formulas for p_1 and p_2. Using these formulas, we will also check that for any dual quaternion q with $|q| \neq 0$; the dual quaternion $q/|q|$ is indeed a unit dual quaternion.

To simplify $q/|q|$, let $q = q_1 + q_2\varepsilon$. Then by Equation (2.3)

$$|q|^2 = |q_1|^2 O + 2(q_1 \cdot q_2)O\varepsilon.$$

But we need to divide q by $|q|$ not by $|q|^2$. Fortunately, it is rather straightforward to take the square root of a dual real number (see Exercise 1). In fact, it is easy to verify that

$$|q| = |q_1|O + \frac{q_1 \cdot q_2}{|q_1|} O\varepsilon \text{ (see Exercise 2).}$$

Therefore

$$\frac{q}{|q|} = \frac{q_1 + q_2\varepsilon}{|q_1|O + \frac{q_1 \cdot q_2}{|q_1|} O\varepsilon} = \left(\frac{q_1 + q_2\varepsilon}{|q_1|O + \frac{q_1 \cdot q_2}{|q_1|} O\varepsilon} \right) \left(\frac{|q_1|O - \frac{q_1 \cdot q_2}{|q_1|} O\varepsilon}{|q_1|O - \frac{q_1 \cdot q_2}{|q_1|} O\varepsilon} \right)$$

$$= \frac{|q_1|q_1 O + \left(|q_1|q_2 - \frac{(q_1 \cdot q_2)}{|q_1|} q_1 \right)\varepsilon}{|q_1|^2},$$

$$\frac{q}{|q|} = \frac{|q_1|^2 q_1 + \left(|q_1|^2 q_2 - (q_1 \cdot q_2)q_1 \right)\varepsilon}{|q_1|^3}.$$

Now we can easily verify that $q/|q|$ is indeed a unit dual quaternion, since

$$\left| \frac{|q_1|^2 q_1}{|q_1|^3} \right| = \frac{|q_1|^3}{|q_1|^3} = 1$$

$$\left(|q_1|^2 q_1 \right) \cdot \left(\left(|q_1|^2 q_2 - (q_1 \cdot q_2)q_1 \right) \right) = |q_1|^4 (q_1 \cdot q_2) - |q_1|^2 (q_1 \cdot q_2)(q_1 \cdot q_1) = 0.$$

The fact that $q/|q|$ is a unit dual quaternion also follows directly from Exercise 2 of Section 1.2.3.

Exercises

1. Let $d = a + b\varepsilon$ be a dual real number with $a \neq 0$. Show that
$$\sqrt{d} = \sqrt{a} + \frac{b}{2\sqrt{a}}\varepsilon.$$

2. Suppose that $|q| = |q_1|O + \dfrac{q_1 \cdot q_2}{|q_1|}\varepsilon$. Show that $|q|^2 = |q_1|^2 O + 2(q_1 \cdot q_2)\varepsilon$.

1.6

Screw Linear Interpolation (ScLERP)

One of the advantages of using unit quaternions instead of rotation matrices to represent rotations in 3-dimensions is that we can use spherical linear interpolation (SLERP) [9, Chapter 9, Section 9.3] to interpolate between two unit quaternions. Since every rigid motion can be represented by a screw transformation (Corollary 4.7), we shall use screw linear interpolation (ScLERP) to interpolate between two rigid motions.

1.6.1 Spherical Linear Interpolation (SLERP) Revisited

To develop a formula for ScLERP, we shall try to mimic the formula we already know for SLERP. Recall from [8, Chapter 15, Section 15.3.5] that for two unit quaternions q_0, q_1

$$SLERP(q_0, q_1, t) = \frac{\sin((1-t)\theta)}{\sin(\theta)} q_0 + \frac{\sin(t\theta)}{\sin(\theta)} q_1, \tag{6.1}$$

where θ is the angle between the unit quaternions q_0 and q_1. Thus

$$SLERP(q_0, q_1, 0) = q_0$$

$$SLERP(q_0, q_1, 1) = q_1,$$

so *SLERP* interpolates between q_0 and q_1; and, most importantly, for all values of t this interpolant is a unit quaternion. We would like to extend this formula for SLERP to interpolate between two unit dual quaternions that represent screw transformations.

Consider then two unit dual quaternions

$$p = p_0 + p_1\varepsilon \quad |p|^2 = O \Rightarrow |p_0|^2 = O, \; p_0 \cdot p_1 = 0$$

$$q = q_0 + q_1\varepsilon \quad |q|^2 = O \Rightarrow |q_0|^2 = O, \; q_0 \cdot q_1 = 0.$$

DOI: 10.1201/9781003398141-7

We could use SLERP to interpolate between the unit quaternions p_0 and q_0, but it is not so clear how to interpolate between p_1 and q_1 so that the interpolant for p_1 and q_1 always remains orthogonal to the interpolant between p_0 and q_0. However, we need this orthogonality condition to ensure that every value of ScLERP is a unit dual quaternion and therefore represents a screw transformation.

We begin this investigation then by introducing a new form for SLERP, which will be easier to mimic when we try to develop a formula for ScLERP (see Section 1.6.3). Let q_0, q_1 be two unit quaternions and consider the quaternion $q_1 q_0^*$. Since q_0 and q_1 are unit quaternions, $q_1 q_0^*$ is also a unit quaternion. Moreover, if the angle between q_0 and q_1 is θ, then the angle between O and $q_1 q_0^*$ is also θ because multiplication by the unit quaternion $q_0^* = q_0^{-1}$ preserves angles. Hence in the quaternion algebra there is a unit vector N such that $q_1 q_0^*$ lies in the ON-plane and

$$q_1 q_0^* = \cos(\theta)O + \sin(\theta)N. \tag{6.2}$$

Multiplying both sides of this equation on the right by q_0 yields

$$q_1 = \cos(\theta)q_0 + \sin(\theta)Nq_0. \tag{6.3}$$

Notice too that since q_0 is a unit quaternion, multiplication by q_0 also preserves lengths and angles. Therefore

$$|N| = 1 \Rightarrow |Nq_0| = 1$$

$$O \perp N \Rightarrow q_0 \perp Nq_0.$$

Now we can set

$$SLERP(q_0, q_1, t) = \cos(t\theta)q_0 + \sin(t\theta)Nq_0, \tag{6.4}$$

since

$$SLERP(q_0, q_1, 0) = q_0$$

$$SLERP(q_0, q_1, 1) = \cos(\theta)q_0 + \sin(\theta)Nq_0 = q_1,$$

and for all values of t this interpolant is a unit quaternion because the length of the product is the product of the lengths (see too Exercise 1). Notice that since q_0, q_1 and θ are known, we can easily solve for the vector N from Equation (6.2). Moreover, it is straightforward to show directly that Equation (6.4) is equivalent to Equation (6.1) (see Exercise 2).

Exercises

1. Let θ be the angle between two unit quaternions q_0 and q_1, and let N be the unit vector defined by Equation (6.2). Show that for all values of t, the expression

$$SLERP(q_0, q_1, t) = \cos(t\theta)q_0 + \sin(t\theta)Nq_0$$

represents a unit quaternion.

2. Let θ be the angle between two unit quaternions q_0 and q_1.
   a. Using Equation (6.3), show that:

   i. $Nq_0 = \dfrac{q_1 - \cos(\theta)q_0}{\sin(\theta)}$

   ii. $\cos(t\theta)q_0 + \sin(t\theta)Nq_0 = \dfrac{\sin((1-t)\theta)}{\sin(\theta)}q_0 + \dfrac{\sin(t\theta)}{\sin(\theta)}q_1.$

   b. Conclude from part *a* that the two formulas for SLERP in Equations (6.1) and (6.4) are equivalent.

1.6.2 The Trigonometric Form of the Screw Transformation

Every rotation in 3-dimensions can be represented as a unit quaternion in trigonometric form:

$$q = q(N, \theta) = \cos(\theta)O + \sin(\theta)N,$$

where 2θ is the angle of rotation and N is a unit (normal) vector parallel to the axis of rotation. In order to extend our construction of the function SLERP from rotations to rigid motions, we are going to show that every screw transformation can be written in a similar trigonometric form using unit dual quaternions by setting

$$S = S(N, \theta) = \cos(\theta_1 + \theta_2 \varepsilon)O + \sin(\theta_1 + \theta_2 \varepsilon)(N_1 + N_2 \varepsilon). \qquad (6.5)$$

Here, however, the angle $\theta = \theta_1 + \theta_2 \varepsilon$ is a dual real number and the unit (normal) vector $N = N_1 + N_2 \varepsilon$ has dual real components.

To understand what this formula actually means, we need to evaluate the trigonometric functions at dual real angles. Using the usual formulas for the sine and cosine of the sum of two angles yields

$$\cos(\theta_1 + \theta_2\varepsilon) = \cos(\theta_1)\cos(\theta_2\varepsilon) - \sin(\theta_1)\sin(\theta_2\varepsilon) \tag{6.6}$$

$$\sin(\theta_1 + \theta_2\varepsilon) = \sin(\theta_1)\cos(\theta_2\varepsilon) + \cos(\theta_1)\sin(\theta_2\varepsilon). \tag{6.7}$$

Now applying the Taylor expansion of $\sin(\theta_2\varepsilon)$ and $\cos(\theta_2\varepsilon)$ and recalling that $\varepsilon^2 = 0$, we get

$$\cos(\theta_2\varepsilon) = 1 \text{ and } \sin(\theta_2\varepsilon) = \theta_2\varepsilon,$$

since all the higher order terms in these Taylor expansions are zero. Substituting these values into Equations (6.6) and (6.7), we find that

$$\cos(\theta_1 + \theta_2\varepsilon) = \cos(\theta_1) - \theta_2\sin(\theta_1)\varepsilon \tag{6.8}$$

$$\sin(\theta_1 + \theta_2\varepsilon) = \sin(\theta_1) + \theta_2\cos(\theta_1)\varepsilon. \tag{6.9}$$

Returning to Equation (6.5), we see now using Equations (6.8) and (6.9) that

$$
\begin{aligned}
S(N,\theta) &= \cos(\theta_1 + \theta_2\varepsilon)O + \sin(\theta_1 + \theta_2\varepsilon)(N_1 + N_2\varepsilon) \\
&= \left(\cos(\theta_1)O + \sin(\theta_1)N_1\right) \\
&\quad + \left(-\theta_2\sin(\theta_1)O + \theta_2\cos(\theta_1)N_1 + \sin(\theta_1)N_2\right)\varepsilon.
\end{aligned}
\tag{6.10}
$$

This formula looks remarkably like the formula in Theorem 4.4 for a screw transformation:

$$
\begin{aligned}
S(N',2\theta,C,2d) &= q(N',\theta) + \sin(\theta)\left((C-O)\times N'\right) \\
&\quad + d\left(-\sin(\theta)O + \cos(\theta)N'\right)\varepsilon.
\end{aligned}
\tag{6.11}
$$

To see if this insight about screw transformations is correct, let's pick apart the components of Equation (6.10). Set

$$q_1 = \cos(\theta_1)O + \sin(\theta_1)N_1.$$

$$q_2 = -\theta_2\sin(\theta_1)O + \theta_2\cos(\theta_1)N_1 + \sin(\theta_1)N_2.$$

Then Equation (6.10) becomes

$$S(N,\theta) = q_1 + q_2\varepsilon.$$

If we insist that $S = S(N,\theta)$ represents a rigid motion, then by Equation (2.4) we must have $|S| = |q_1| = 1$. Therefore $|N_1| = 1$, so $q_1 = q(N_1,\theta_1)$ is just a unit quaternion representing a rotation by the angle $2\theta_1$ around an axis parallel to the unit (normal) vector N_1. Moreover, again if S represents a rigid motion, then again by Equation (2.4) since $|S| = 1$, we must have

$$0 = q_1 \cdot q_2 = \sin^2(\theta_1)(N_1 \cdot N_2) \Rightarrow N_1 \cdot N_2 = 0 \Rightarrow N_2 \perp N_1.$$

Summarizing: with these constraints, the trigonometric form of a screw transformation is

$$S(N,\theta) = \big(\cos(\theta_1)O + \sin(\theta_1)N_1\big)$$
$$+ \big(-\theta_2\sin(\theta_1)O + \theta_2\cos(\theta_1)N_1 + \sin(\theta_1)N_2\big)\varepsilon \tag{6.12}$$

where by definition (Equation 6.5) $\theta = \theta_1 + \theta_2\varepsilon$, $N = N_1 + N_2\varepsilon$. Moreover, we can now show that $S(N,\theta)$ is the screw transformation where:

    i. $N_1 = N'$ is a unit (normal) vector parallel to the axis of rotation
    ii. $2\theta_1$ is the angle of rotation
    iii. $N_2 \perp N_1$
    iv. $C = O + (N_1 \times N_2)\varepsilon$ is the center of rotation
    v. $2\theta_2$ is the translation along the axis of rotation parallel to N_1.

To understand why properties i–v hold, let's compare Equations (6.10) and (6.11). Performing this comparison, we see immediately that properties i and ii are now clear, and property iii we have already proved. To derive property iv, observe that since $N_2 \perp N_1$, there exist scalars c_1, c_2, c_3 such that

$$C - O = c_1N_1 + c_2N_2 + c_3N_1 \times N_2 \Rightarrow$$
$$(C - O) \times N_1 = c_2N_2 \times N_1 + c_3(N_1 \times N_2) \times N_1.$$

But $|N_1| = 1$ and $N_2 \perp N_1$, so

$$(N_1 \times N_2) \times N_1 = N_2.$$

Therefore

$$(C - O) \times N_1 = c_2N_2 \times N_1 + c_3N_2.$$

Now comparing Equations (6.10) and (6.11), we see that $c_2 = 0$, $c_3 = 1$, since there is an N_2 term but no $N_2 \times N_1$ term in Equation (6.10). Thus in accord with property iv, we can take the center of rotation to lie at

$$C = O + N_1 \times N_2,$$

so

$$\sin(\theta)((C - O) \times N_1) = \sin(\theta) N_2 \varepsilon.$$

Notice that the component of $C - O$ along N_1 does not matter because we are taking the cross product of $C - O$ with $N' = N_1$. Finally, comparing the remaining terms in Equations (6.10) and (6.11), it follows that $\theta_2 = d$, which gives property v. We conclude then that

$$S(N, \theta) = \cos(\theta_1 + \theta_2 \varepsilon) O + \sin(\theta_1 + \theta_2 \varepsilon)(N_1 + N_2 \varepsilon)$$

is a screw transformation for every unit (normal) vector $N = N_1 + N_2 \varepsilon$ and every value of $\theta = \theta_1 + \theta_2 \varepsilon$, and that every screw transformation can be expressed in this form.

Exercises

1. Using only the Taylor expansions of sine and cosine and without appealing to the identities for the sine and cosine of the sum of two angles, show that:
   a. $\cos(\theta_1 + \theta_2 \varepsilon) = \cos(\theta_1) - \theta_2 \sin(\theta_1)\varepsilon$
   b. $\sin(\theta_1 + \theta_2 \varepsilon) = \sin(\theta_1) + \theta_2 \cos(\theta_1)\varepsilon$.
   (Hint: Show that: $(\theta_1 + \theta_2 \varepsilon)^n = \theta_1^n + n\theta_1^{n-1}\theta_2 \varepsilon$.)

2. Let $S(N, \theta) = \cos(\theta_1 + \theta_2 \varepsilon) O + \sin(\theta_1 + \theta_2 \varepsilon)(N_1 + N_2 \varepsilon)$. Show how to extract θ_1, θ_2 and N_1, N_2 from the expression for $S(N, \theta)$. {Hint: Use Equation (6.10) and compare separately the coefficients of O, N_1, and ε.}

1.6.3 ScLERP

With the trigonometric form of screw transformations now in hand we are finally ready to interpolate between two screw transformations represented by two unit dual quaternions. Let S_0, S_1 be two unit dual quaternions representing two screw transformations. Then $S_1 S_0^*$ is also a unit dual quaternion and also represents a screw transformation because the product of two rigid motions is a rigid motion and therefore a screw transformation. Hence there are angles $\theta = \theta_1 + \theta_2 \varepsilon$ and vectors $N = N_1 + N_2 \varepsilon$ such that $|N_1| = 1$, $N_1 \cdot N_2 = 0$, and

$$S_1 S_0^* = \cos(\theta)O + \sin(\theta)N$$

$$= \cos(\theta_1 + \theta_2\varepsilon)O + \sin(\theta_1 + \theta_2\varepsilon)(N_1 + N_2\varepsilon). \tag{6.13}$$

Multiplying both sides of Equation (6.13) on the right by S_0, we get

$$S_1 = \cos(\theta)S_0 + \sin(\theta)NS_0 = \cos(\theta_1 + \theta_2\varepsilon)S_0 + \sin(\theta_1 + \theta_2\varepsilon)(N_1 + N_2\varepsilon)S_0.$$

Now in analogy with SLERP, we define

$$\begin{aligned} ScLERP(S_0, S_1, t) &= \cos(t\theta)S_0 + \sin(t\theta)NS_0 \\ &= \cos(t\theta_1 + t\theta_2\varepsilon)S_0 \\ &\quad + \sin(t\theta_1 + t\theta_2\varepsilon)(N_1 + N_2\varepsilon)S_0. \end{aligned} \tag{6.14}$$

Then clearly

$$ScLERP(S_0, S_1, 0) = S_0$$

$$ScLERP(S_0, S_1, 1) = \cos(\theta)S_0 + \sin(\theta)NS_0 = S_1.$$

Moreover, crucially $ScLERP(S_0, S_1, t)$ is a screw transformation for all values of t (see Exercise 1). Notice that since S_0, S_1 are known, we can solve for the angle θ and the (normal) vector N from Equation (6.13) (see Section 1.6.2, Exercise 2).

Exercises

1. Let $N = N_1 + N_2\varepsilon$, where N_1, N_2 are dual quaternions representing normal vectors, and let S be a dual quaternion representing a screw transformation.
   a. Show that:
      i. $S \cdot (NS) = 0$
      ii. $N_1 \cdot N_2 = 0 \Rightarrow (N_1 S) \cdot (N_2 S) = 0$
      (Hint: See Equation (2.6).)
   b. Conclude that $ScLERP(S_0, S_1, t)$ represents a screw transformation for all values of t.

2. Using Equations (6.8) and (6.9) to unwind the definition of ScLERP in Equation (6.14), show that:

    a. $ScLERP(S_0, S_1, t) = \left(\cos(t\theta_1)S_0 + \sin(t\theta_1)NS_0\right)$

    $$+ \left(-t\theta_2\sin(t\theta_1)S_0 + t\theta_2\cos(t\theta_1)NS_0\right)\varepsilon$$

    b. $ScLERP(S_0, S_1, t)$

    $$= \left(\cos(t\theta_1)S_0 + \sin(t\theta_1)N_1S_0\right)$$

    $$+ \left(-t\theta_2\sin(t\theta_1)S_0 + t\theta_2\cos(t\theta_1)N_1S_0 + \sin(t\theta_1)N_2S_0\right)\varepsilon.$$

3. Suppose that the rigid motions R_0, R_1 represent rotations around a line through the origin. Show that

    $$ScLERP(R_0, R_1, t) = SLERP(R_0, R_1, t).$$

4. Suppose that the rigid motions R_0, R_1 represent translations. Show that $ScLERP(R_0, R_1, t)$ is just linear interpolation—that is, for two translations show that

    $$ScLERP(R_0, R_1, t) = R_0 + t(R_1 - R_0).$$

5. Suppose that S_0, S_1 are two unit dual quaternions, S_0 representing a rotation and S_1 representing a translation—that is,

    $$S_0 = R = \cos(\phi)O + \sin(\phi)v$$

    $$S_1 = T = O + \frac{1}{2}dt\varepsilon$$

    a. Show that in this case

    $$ScLERP(R, T, \tau) = \cos(\phi + \tau\theta_1)O + \sin(\phi + \tau\theta_1)v$$

    $$+ \left(\sin(\tau\theta_1)\left(\cos(\phi)N_2 + \sin(\phi)N_2v\right) + \tau\theta_2v\right)\varepsilon$$

    where the parameters in Equations (6.13) and (6.14) are chosen so that

    $$\phi + \theta_1 = 0, \quad N_1 = v, \quad \sin(\theta_1)\left(\cos(\phi)N_2 + \sin(\phi)N_2v\right) + \theta_2v = \frac{d}{2}t.$$

    b. Using part *a*, show that $t \parallel v \Rightarrow N_2 = 0$ and $t \perp v \Rightarrow \theta_2 = 0$.

1.7

Perspective and Pseudo-Perspective

Perspective projections are necessary in computer graphics, as in classical art, in order to render realistic images. Therefore, rigid motions are not sufficient for computer graphics. Here we shall show how to use dual quaternions to compute perspective and pseudo-perspective projections. These applications of dual quaternions are relatively new and not available in standard presentations on dual quaternions.

1.7.1 Perspective in the Quaternion Algebra

In the quaternion algebra we can compute perspective projections by sandwiching a vector between two copies of a unit quaternion (see [9, Chapter 6, Section 6.3]). The key trick with quaternions is to use a rotation in the ON-plane to convert some of the distance along the unit normal N to the perspective plane into mass at the origin O. In Section 1.7.3 we shall show how to use sandwiching with unit dual quaternions along with a similar trick to compute perspective projections using dual quaternions. But first we are going to reexamine more closely how perspective projections are computed in the quaternion algebra to understand why we cannot use the identical approach to compute perspective projections in the algebra of dual quaternions.

The quaternions are a subalgebra of the dual quaternions. This embedding allows us to perform rotations in the space of dual quaternions using the same approach as in the quaternion subalgebra. We might hope then to use this quaternion subalgebra to perform perspective projection in the space of dual quaternions. But now there is a problem: the representation of geometry in the space of dual quaternions is different from the representation geometry in the space of quaternions. Vectors v in the quaternion algebra are represented in the algebra of dual quaternions by the expressions $v\varepsilon$ and points $P = O + v_P$ in the quaternion subalgebra are represented in the algebra of dual quaternions by the expressions $P = O + v_P\varepsilon$. Now consider an eye point E and an arbitrary point P in 3-dimensions. The formula for perspective projection in the space of quaternions sandwiches the vector $P - E$ between two copies of a unit quaternion q to generate a

DOI: 10.1201/9781003398141-8

mass-point, where the associated point is located at the perspective projection of the point P from the eye point E onto a perspective plane. But the vectors v in the quaternion algebra are represented in the algebra of dual quaternions by the expressions $v\varepsilon$, and the sandwiching formula

$$q(v\varepsilon)q = (qvq)\varepsilon = (mO+u)\varepsilon$$

is not a mass-point in the space of dual quaternions, since in the space of dual quaternions mass-points all have the form $mO+u\varepsilon$. Thus, when using the dual quaternion representation for points and vectors, we cannot simply apply sandwiching with unit quaternions representing rotations to compute perspective projections. Rather we shall apply sandwiching with the dual quaternions representing translations together with duality. To motivate this approach, however, we first need to examine how rotations and translations behave with respect to duality.

1.7.2 Rotation, Translation, and Duality

Rotation using quaternions still works in the space of dual quaternions when we replace points and planes by their duals. Consider a point $P = O + v_P\varepsilon$. Then

$$qP^{\#}q^{*} = q(O\varepsilon + v_P)q^{*} = qO\varepsilon q^{*} + qv_Pq^{*} = O\varepsilon + qv_Pq^{*}.$$

Thus, sandwiching between the quaternion

$$q(N,\theta/2) = \cos(\theta/2)O + \sin(\theta/2)N$$

and its conjugate rotates the planes dual to points and vectors by the angle θ around the line through the origin O parallel to the unit normal vector N; nothing really changes for rotation around lines through the origin when we replace the origin O by the plane at infinity $O\varepsilon$ in our dual representation. Similarly, for a plane $\pi = v_{\pi} + dO\varepsilon$, the sandwiching map

$$\pi^{\#} \mapsto q\pi^{\#}q^{*} = (qv_{\pi}q^{*})\varepsilon + dO$$

rotates the mass-point dual to the plane π by the angle θ around the line through the origin O parallel to the unit normal vector N.

But even with duality we still cannot use the quaternion subalgebra to compute perspective projection by sandwiching the dual of a vector between two copies of a unit quaternion because for any vector $v\varepsilon$ the expression

$$q(v\varepsilon)^{\#} q = qvq = mO + u$$

is still not a mass-point in the space of dual quaternion: in the space of dual quaternions, mass-points have the form $mO + u\varepsilon$.

But now we have another alternative. The subalgebra of unit quaternions is a collection of isometries in R^8, isometries which can be used to rotate points and vectors in 3-dimensions. There are, however, other isometries in R^8, the isometries that can be used to translate points in 3-dimensions. Let's try these isometries.

First recall that sandwiching with the maps $T = T\left(N, \dfrac{d}{2}\right) = O + \dfrac{d}{2} N\varepsilon$ translate points in the direction of the unit vector $N\varepsilon$ by the distance d.

In particular, the origin O is mapped to the point $O + dN\varepsilon$. But notice that sandwiching with T does not alter the dual to the origin, since

$$T\left(O^{\#}\right)T = T(O\varepsilon)T = \left(O + \dfrac{d}{2} N\varepsilon\right)O\varepsilon\left(O + \dfrac{d}{2} N\varepsilon\right) = O\varepsilon = O^{\#}.$$

This result makes sense because $O^{\#} = O\varepsilon$ is the point at infinity, and the point at infinity is unaffected by translation.

Vectors too are unaffected by translation. Let's see, however, what these translation maps now do to the duals of vectors. For future computational convenience let's replace d by $-1/d$ and set

$$T = T\left(N, -\dfrac{1}{2d}\right) = O - \dfrac{1}{2d} N\varepsilon.$$

Now for any vector $v\varepsilon$, we have the sandwiching formula

$$T(v\varepsilon)^{\#}T = \left(O - \dfrac{1}{2d} N\varepsilon\right)v\left(O - \dfrac{1}{2d} N\varepsilon\right) = \left(v - \dfrac{1}{2d} Nv\varepsilon\right)\left(O - \dfrac{1}{2d} N\varepsilon\right)$$

$$= v - \dfrac{1}{2d} Nv\varepsilon - \dfrac{1}{2d} vN\varepsilon = v - \dfrac{1}{d}\left(\dfrac{Nv + vN}{2}\right)\varepsilon.$$

But in the quaternion algebra $-\dfrac{Nv + vN}{2} = (N \cdot v)O$, so

$$T(v\varepsilon)^{\#}T = v + \dfrac{1}{d}(N \cdot v)O\varepsilon. \tag{7.1}$$

Here we have simply repeated the computation in Section 1.4.4 for how translation behaves on a normal vector to a plane. This result also makes sense since while vectors are unaffected by translation, planes, and in particular their distances to the origin, are affected by translation.

Thus, sandwiching the dual $v = (v\varepsilon)^{\#}$ to a vector $v\varepsilon$ between two copies of a translation T introduces a constant multiple along the dual $O^{\#} = O\varepsilon$ to the origin O; taking the dual yet again will introduce mass at the origin, just the device we need for computing perspective projection (see [8, Chapter 6]). These observations lead to a method for computing perspective projections in the space of dual quaternions using translations composed with duality, but before we can proceed with this approach, we need an explicit formula for perspective projection.

1.7.3 Perspective Projection

Proposition 7.1 (Explicit Formula for Perspective Projection)

Let

- $E = O = eye\ point$
- $N\varepsilon = a\ unit\ vector\ pointing\ from\ the\ eye\ point\ E\ to\ the\ perspective\ plane\ \pi$

- $\pi = O\varepsilon - \dfrac{1}{d}N = the\ perspective\ plane\ at\ a\ distance\ d\ from\ the\ eye\ point\ E\ perpendicular\ to\ the\ unit\ normal\ vector\ N.$
- $P = O + v_P\varepsilon = an\ arbitrary\ point$

Then the mass-point

- $P' = \dfrac{1}{d}(N \cdot v_P)O + v_P\varepsilon$

is located at the perspective projection of the point P from the eye point E onto the plane π.

Proof: To show that the mass-point P' is located at the perspective projection of the point P from the eye point E onto the plane π, we need to show that P' is located at the intersection of the plane π with the line from the eye point E to the point P. Now observe that the mass-point $P' = \dfrac{1}{d}(N \cdot v_P)O + v_P\varepsilon$ is located at $O + \dfrac{d}{N \cdot v_P}v_P\varepsilon$, which is a point on the line $O + tv_P \varepsilon$ from the eye point $E = O$ to the point $P = O + v_P \varepsilon$. Moreover, the mass-point P' lies on the plane π since

$$P' \cdot \pi = \left(\frac{1}{d}(N \cdot v_P)O + v_P \varepsilon \right) \cdot \left(O\varepsilon - \frac{1}{d}N \right) = \frac{1}{d}((N \cdot v_P) - (v_P \cdot N))O\varepsilon = 0. \ \Diamond$$

Theorem 7.2 (Perspective Projection)

Let

- $E = O = eye\ point$
- $N\varepsilon = a\ unit\ vector\ pointing\ from\ the\ eye\ point\ E\ to\ the\ perspective$
  plane π
- $\pi = O\varepsilon - \frac{1}{d}N = the\ perspective\ plane\ at\ a\ distance\ d\ from\ the\ eye\ point$

  $E = O\ perpendicular\ to\ the\ unit\ normal\ vector\ N.$
- $T = O - \frac{1}{2d}N\varepsilon = a\ translation\ operator$

Then $\left(T(P - E)^{\#} T \right)^{\#}$ *is a mass-point* (mP', m) *where:*

- *the point P' is located at the perspective projection of the point P from the eye point $E = O$ onto the plane π;*
- *the mass $m = s/d$ where s is the distance of the point P from the plane through the eye point E perpendicular to the unit normal vector N, and d is the fixed distance from the eye point E to the perspective plane π.*

Proof: Let $P = O + v_P \varepsilon$. Then by Equation (7.1)

$$T(P - E)^{\#} T = T(v_P \varepsilon)^{\#} T = v_P + \frac{1}{d}(N \cdot v_P)O\varepsilon.$$

Taking the dual yields

$$\left(T(P - E)^{\#} T \right)^{\#} = \left(v_P + \frac{1}{d}(N \cdot v_P)O\varepsilon \right)^{\#} = \frac{1}{d}(N \cdot v_P)O + v_P \varepsilon.$$

Therefore, by Proposition 7.1 $\left(T(P - E)^{\#} T \right)^{\#}$ is located at the perspective projection of the point P from the eye point E onto the plane π. Moreover, $N \cdot v_P = $ the distance s of the point P from the plane through the eye point $E = O$ perpendicular to the unit normal vector N (see Figure 7.1).

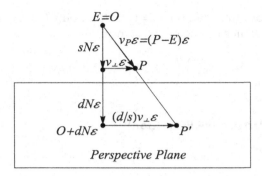

FIGURE 7.1
Perspective projection.

Thus, the mass $m = \dfrac{1}{d}(N \cdot v_P)$ is equal to the distance $N \cdot v_P$ of the point P from the plane through the eye point $E = O$ perpendicular to the unit normal vector N divided by the fixed distance d from the eye point E to the perspective plane π. $\Diamond$

Notice that $\left(T(P-E)^{\#}T\right)^{\#}$ computes more than just the location of the perspective projection, $\left(T(P-E)^{\#}T\right)^{\#}$ also computes a mass. The mass $m = s/d$ is a constant $1/d$ times the distance s from the point P to the plane through the eye point E parallel to the plane of projection π. Therefore, we can use this sandwiching formula for perspective projection, just like we use the sandwiching formula for perspective projection with quaternions in [9, Chapter 6] to detect hidden surfaces: if two points project to the same point, then the smaller the mass, the closer the surface point is to the eye. Points further from the plane of the eye are hidden by points closer to the plane of the eye. Moreover, unlike the sandwiching formula for perspective projection using unit quaternions in [9, Chapter 6], there is no constraint here on the distance d from the eye point E to the perspective plane π.

One of the reasons that translation together with duality can be used to compute perspective projection is that translation does not behave symmetrically with respect to vectors and normal vectors. Consider a translation $T = O - \dfrac{1}{2d}N\varepsilon$. Recall that

$$T(v\varepsilon)T = v\varepsilon \qquad\qquad\qquad\text{(vectors)}$$

$$TvT = v - \frac{1}{d}\left(\frac{Nv + vN}{2}\right)O\varepsilon \quad \text{(normal vectors)}$$

so

$$T(v\varepsilon)^{\#}T \neq \left(T(v\varepsilon)T\right)^{\#}$$

$$\left(T(v\varepsilon)^{\#}T\right)^{\#} \neq T(v\varepsilon)T.$$

Thus, translation and duality do not commute: duality followed by translation followed by duality introduces mass. Contrast this behavior with a rotation represented by a unit quaternion $q = \cos(\theta/2)O + \sin(\theta/2)N$. It is straightforward to verify (see Exercise 3) that

$$q(v\varepsilon)q^{+} = \left(qvq^{+}\right)\varepsilon$$

$$\left(q(v\varepsilon)q^{+}\right)^{\#} = qvq^{+} = q(v\varepsilon)^{\#}q^{+} \qquad (7.2)$$

Thus, rotation and duality commute: rotating and then taking the dual is the same as taking the dual first and then rotating.

Since rotation and duality commute, we can compose rotation and perspective projection in the same way that we compose rotation and translation: by taking the product of their representations as dual quaternions. Indeed, by Equation (7.2):

Rotation followed by Perspective:

$$P \mapsto \left(T\left(q(P-E)q^{+}\right)^{\#}T^{+}\right)^{\#} = \left((Tq)(P-E)^{\#}(Tq)^{+}\right)^{\#}.$$

Perspective followed by Rotation:

$$P \mapsto q\left(T(P-E)^{\#}T\right)^{\#}q^{+} = \left((qT)(P-E)^{\#}(qT)^{+}\right)^{\#}.$$

Note that here the rotation q and the perspective map T are both with respect to the eye point at the origin—that is, the rotation is around a line through the origin.

Exercises

1. Here we provide an alternative proof of Theorem 7.2. For this exercise, we adopt the notation of Theorem 7.2 for N, E, π, T, P, v_P.

   a. Show that there is a scalar s and a vector $v_{\perp}\varepsilon$ perpendicular to the vector $N\varepsilon$ such that

      i. $v_P\varepsilon = sN\varepsilon + v_{\perp}\varepsilon$ (see Figure 7.1)

      ii. $T(P-E)^{\#}T = sN + v_{\perp} + \dfrac{s}{d}O\varepsilon$ (Hint: Use Equation 7.1.)

    b. Show that the point $P' = O + \left(dN + \dfrac{d}{s} v_\perp \right) \varepsilon$

       i. lies on the plane π

      ii. lies on the line from the origin O to the point P

    c. Conclude from parts a and b that:

       i. the point P' is the perspective projection of the point P from the eye point $E = O$ to the plane π;

      ii. the mass-point $\left(T(P - E)^\# T \right)^\#$ is located at the perspective projection P' of the point P from the eye point $E = O$ to the plane π.

2. Here we provide yet another proof of Theorem 7.2. For this exercise, we adopt the notation of Exercise 1 for $T, E, P, P', \pi, v_\perp, s, d$.

    a. Use similar triangles to show that P' is the perspective projection of the point P from the eye point $E = O$ to the plane π (see Figure 7.1).

    b. Conclude from part a that $\left(T(P - E)^\# T \right)^\#$ is located at the perspective projection P' of the point P from the eye point $E = O$ to the plane π.

3. Show that duality and rotation around a line through the origin commute.

4. Show that duality and reflection in a plane through the origin commute.

5. In this exercise we shall show how to compute perspective projection using the rotation operator together with a novel form of duality. Define a new duality operator $x^\$$ by setting

   • $i^\$ = i\varepsilon$ and $(i\varepsilon)^\$ = i$

   • $j^\$ = j\varepsilon$ and $(j\varepsilon)^\$ = j$

   • $k^\$ = k\varepsilon$ and $(k\varepsilon)^\$ = k$

   • $O^\$ = O$ and $(O\varepsilon)^\$ = O\varepsilon$

and extending to arbitrary dual quaternions by linearity. On the basis elements, $x^\$$ differs from $x^\#$ only on O and $O\varepsilon$. Unlike the operator $x^\#$, which maps points and vectors to planes, and maps planes to points or vectors, the operator $x^\$$ maps points to quaternions and maps planes to quaternions multiplied by ε. Notice, however, it is still true that $\left(x^\$ \right)^\$ = x$. Now to perform perspective projection using the rotation operator, let

   • $N\varepsilon = $ a unit vector

   • $E = O - N\varepsilon = $ eye point

- $\pi = N =$ perspective plane through the origin perpendicular to the normal vector N

- $q = \cos(\pi/4) - \sin(\pi/4)N =$ rotation by $-\pi/2$ around the line through the origin O parallel to the unit normal vector N
  Show that $\left(q(P-E)^\$ q\right)^\$$ is a mass-point (mP', m), where

- the point P' is located at the perspective projection of the point P from the eye point E onto the plane π;

- the mass m is equal to the distance d of the point P from the plane through the eye point E perpendicular to the unit normal vector N.
  (Hint: See [9, Chapter 6, Theorem 6.8].)

6. Repeat Exercise 5 this time with

- $N\varepsilon =$ a unit vector

- $E = O + \left(\cot(\theta) - \csc(\theta)\right)N\varepsilon =$ eye point

- $\pi = \cot(\theta)O\varepsilon - N =$ perspective plane perpendicular to the unit normal vector N at a distance $\csc(\theta)$ from the eye point E

- $q = \cos(\theta/2) - \sin(\theta/2)N =$ rotation by $-\theta$ around the line through the origin O parallel to the unit normal vector N

  Show that $\left(q(P-E)^\$ q\right)^\$$ is a mass-point (mP', m), where

- the point P' is located at the perspective projection of the point P from the eye point E onto the plane π;

- the mass $m = d\sin(\theta)$, where d is the distance of the point P from the plane through the eye point E perpendicular to the unit normal vector N.
  (Hint: Apply [9, Chapter 6, Theorem 6.9].)

1.7.4 Pseudo-Perspective

Pseudo-perspective maps the eye point to a vector (i.e., a point at infinity) and a viewing frustum to a rectangular box [8,14] (see Figure 7.2). Pseudo-perspective is used in computer graphics for three purposes: (i) to speed up clipping algorithms by replacing the viewing frustum with a rectangular box; (ii) to simplify projections by replacing perspective projection with orthogonal projection; and (iii) to expedite scan converting triangles with hidden surface removal. Thus, pseudo-perspective is often more important in computer graphics than ordinary perspective projection.

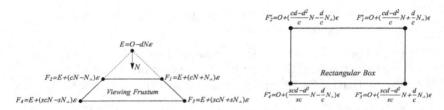

FIGURE 7.2
Schematic mapping of a viewing frustum into a rectangular box. The points F_i^* are the images of the points F_i for $i = 1, \ldots, 4$.

When we represent points and vectors using quaternions, we can also use unit quaternions to compute pseudo-perspective (see Exercise 1). But this approach breaks down in the space of dual quaternions for the same reason that unit quaternions cannot be used to compute perspective projection. Thus, once again to compute pseudo-perspective using dual quaternions, we turn to duality and we invoke the transformations $T\left(N, -\dfrac{1}{2d}\right) = O - \dfrac{1}{2d}N\varepsilon$, which we use to compute perspective projection.

Theorem 7.3 (Pseudo-Perspective)

Let

- $E = O - dN\varepsilon = eye\ point$
- $N\varepsilon = a\ unit\ vector$
- $T = O - \dfrac{1}{2d}N\varepsilon = a\ translation\ operator$

Then the map $P \to \left(TP^\# T\right)^\#$ *transforms the eye point E to the point at infinity in the direction* $-N\varepsilon$ *and transforms a viewing frustum to a rectangular box.*
  Proof: By Equation (7.1)

$$TvT = v + \frac{1}{d}(N \cdot v)O\varepsilon.$$

Therefore

$$v_{\parallel} = sN \Rightarrow Tv_{\parallel}T = sN + \frac{s}{d}O\varepsilon \qquad (7.3)$$

$$v_{\perp} \perp N \Rightarrow Tv_{\perp}T = v_{\perp} \qquad (7.4)$$

Hence by linearity and Equation (7.3)

$$TE^{\#}T = T(O\varepsilon - dN)T = T(O\varepsilon)T + T(-dN)T = O\varepsilon - (dN + O\varepsilon) = -dN$$

so $E \rightarrow \left(TE^{\#}T\right)^{\#}$ maps the eye point E to a point at infinity in the direction $-N\varepsilon$.

Now let $N_{\perp}$ denote any vector perpendicular to N and consider four points $F_1, \ldots, F_4$ on a viewing frustum (see Figure 7.2). Again, by linearity and Equations (7.3) and (7.4):

$$F_1 = E + cN\varepsilon + N_{\perp}\varepsilon \rightarrow \left(TF_1^{\#}T\right)^{\#} = -dN\varepsilon + cN\varepsilon + \frac{c}{d}O + N_{\perp}\varepsilon$$

$$\equiv O + \left(\frac{cd - d^2}{c}N + \frac{d}{c}N_{\perp}\right)\varepsilon$$

$$F_2 = E + cN\varepsilon - N_{\perp}\varepsilon \rightarrow \left(TF_2^{\#}T\right)^{\#} = -dN\varepsilon + cN\varepsilon + \frac{c}{d}O - N_{\perp}\varepsilon$$

$$\equiv O + \left(\frac{cd - d^2}{c}N - \frac{d}{c}N_{\perp}\right)\varepsilon$$

$$F_3 = E + scN\varepsilon + sN_{\perp}\varepsilon \rightarrow \left(TF_3^{\#}T\right)^{\#} = -dN\varepsilon + scN\varepsilon + \frac{sc}{d}O + sN_{\perp}\varepsilon$$

$$\equiv O + \left(\frac{scd - d^2}{sc}N + \frac{d}{c}N_{\perp}\right)\varepsilon$$

$$F_4 = E + scN\varepsilon - sN_{\perp}\varepsilon \rightarrow \left(TF_4^{\#}T\right)^{\#} = -dN\varepsilon + scN\varepsilon + \frac{sc}{d}O - sN_{\perp}\varepsilon.$$

$$\equiv O + \left(\frac{scd - d^2}{sc}N - \frac{d}{c}N_{\perp}\right)\varepsilon$$

Since the factor d/c is independent of the scale factor s, the map $P \rightarrow \left(TP^{\#}T\right)^{\#}$ transforms the viewing frustum into a rectangular box (see Figure 7.2). ◊

In our theorems for perspective and pseudo-perspective, we have located the eye point E at a special canonical position for which the results are especially easy to compute. To deal with arbitrary eye points E', we can simply translate the entire scene—points and planes—by the vector $v = E - E'$ before performing perspective or pseudo-perspective and then after performing perspective or pseudo-perspective with the eye point in canonical position, translate the resulting scene by $-v = E' - E$ to move the outcome back to the required position. For perspective, the vector $P - E$ is unaffected by translation, so we need only perform the final translation and can dispense altogether with the initial translation.

Exercise

1. **Pseudo-Perspective Using Quaternions**: Recall that in the space of quaternions, the expression $P = O + v_p$ represents a point in 3-dimensions. Let

   $E = O - \cot(\theta)N = eye\ point$

   $N = a\ unit\ vector$

   $q = q(N, \theta/2) = \cos(\theta/2)O + \sin(\theta/2)N$

   Show that:

   a. the map $P \rightarrow qPq$ correctly computes pseudo-perspective in the space of quaternions by transforming the eye point E to a vector parallel to N and transforming a viewing frustum to a rectangular box.

   b. the map $P \rightarrow qPq$ does not correctly compute pseudo-perspective in the space of dual quaternions.

   c. the map $P \rightarrow \left(qP^{\$}q\right)^{\$}$, where $x^{\$}$ is the duality operator defined in Section 1.7.3, Exercise 5, $E = O - \cot(\theta)N\varepsilon$, and $P = O + v_p\varepsilon$ is a point in the space of dual quaternions, correctly computes pseudo-perspective in the space of dual quaternions.

1.8

Visualizing Quaternions and Dual Quaternions

To study quaternions, we can use two orthogonal planes in 4-dimensions to visualize simultaneously all 4-dimensions (see [9, Chapter 3, Section 3.3] and Figure 8.1). Using this pair of orthogonal planes, we are able to understand visually how each of the sandwiching maps $x \mapsto qxq^*$ and $x \mapsto qxq$, where q is a unit quaternion, represents a simple rotation in one of these two orthogonal planes (see [9, Chapter 6]).

To understand dual quaternions, we need four mutually orthogonal planes to visualize simultaneously all 8-dimensions (see Figure 8.2). We group these four planes into two pairs of orthogonal planes: one pair to represent points and vectors and one pair to represent the duals of points and vectors—that is, to represent planes and normal vectors in the space of dual quaternions.

One key fact about the pair of planes representation for quaternions is that the plane in 4-dimensions containing the vector corresponding to the origin O in 3-dimensions is special: algebraically this plane is isomorphic to the complex plane (see Figure 8.3). The corresponding fundamental fact about the four planes representation for dual quaternions is that the plane in 8-dimensions containing the vector corresponding to the origin O in 3-dimensions is also special: algebraically this plane is isomorphic to the plane of dual real numbers (see Figure 8.4).

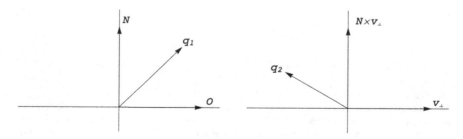

FIGURE 8.1

Two orthogonal planes in 4-dimensions: N is a unit vector in 3-dimensions and $v_\perp$ is a unit vector perpendicular to N. We can visualize any quaternion q by its projection q_1, q_2 into these two orthogonal planes.

DOI: 10.1201/9781003398141-9

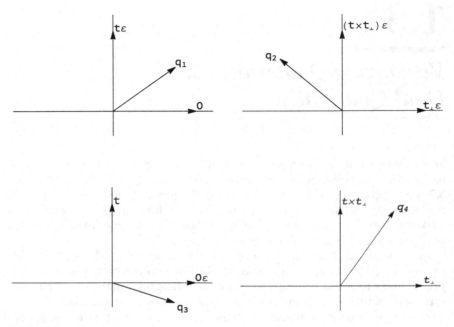

FIGURE 8.2
Four mutually orthogonal planes in 8-dimensions: t is a unit vector in 3-dimensions and $t_\perp$ is a unit vector perpendicular to t. We can visualize any dual quaternion q by its projections q_1, q_2, q_3, q_4 into these four orthogonal planes. The top two planes are used to represent points and vectors; the bottom two planes are used to represent planes and normal vectors in the space of dual quaternions.

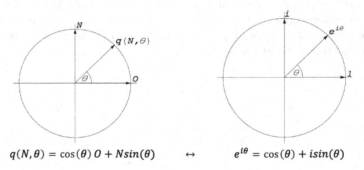

$$q(N, \theta) = \cos(\theta)\, O + N\sin(\theta) \qquad \leftrightarrow \qquad e^{i\theta} = \cos(\theta) + i\sin(\theta)$$

FIGURE 8.3
For quaternions the plane (left) containing the vector in 4-dimensions corresponding to the origin O in 3-dimensions is isomorphic to the complex plane (right): N is a unit vector in 3-dimensions, so in quaternion multiplication $N^2 = -O$. Also, O is the identity for quaternion multiplication, so $O^2 = 1$ and $ON = NO = N$. These identities mimic the algebra of the complex numbers, where $i^2 = -1$, and 1 is the identity for complex multiplication. The quaternion $q(N, \theta)$ corresponds to the complex number $e^{i\theta}$; both represent rotations in their respective planes by the angle θ.

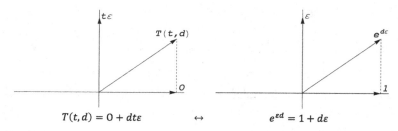

$$T(t,d) = O + dt\varepsilon \qquad\qquad \leftrightarrow \qquad\qquad e^{\varepsilon d} = 1 + d\varepsilon$$

FIGURE 8.4

For dual quaternions the plane (left) containing the vector in 8-dimensions corresponding to the origin O in 3-dimensions is isomorphic to the plane of dual real numbers (right). The vector t is a unit vector in 3-dimensions, but $\varepsilon^2 = 0$, so $(t\varepsilon)^2 = 0$. The dual quaternion $T(t,d) = O + dt\varepsilon$ corresponds to the dual real number $e^{\varepsilon d} = 1 + d\varepsilon$; both represent shears in their respective planes in the vertical direction. In the space of dual quaternions, these shears are isometries because the direction along $t\varepsilon$ carries no length, and the angle between O and $t\varepsilon$ is unchanged by what appears to be a shear in the $O(t\varepsilon)$-plane.

We can use this model of four mutually orthogonal planes to visualize two important effects in 8-dimensions: the effect of sandwiching with the isometries that represent rotations (see Figure 8.5) and the effect of sandwiching with the isometries that represent translations (see Figure 8.6). Notice that a shear in the 4-dimensional representation for points and vectors represents a translation in 3-dimensions, since this shear in 4-dimensions relocates the origin in 3-dimensions. Also, a shear in the 4-dimensional representation for the duals of points and vectors can be used together with duality to represent perspective projection in 3-dimensions, since this shear in 4-dimensions introduces some mass at the dual $O\varepsilon$ to the origin O. For more about the relationship between shears in 4-dimensions and transformations in 3-dimensions, see [7].

A word of caution: we need to be careful here how we interpret these pictures because unlike the metric for quaternions, the metric for dual quaternions is not the standard Euclidean metric: directions along multiples of ε carry no length. Indeed, if $q = p\varepsilon$, then

$$|q|^2 = qq^* = (p\varepsilon)(p\varepsilon)^* = |p|^2\,\varepsilon^2 = 0.$$

Moreover, by Equation (2.2) vectors orthogonal to these ε-directions remain orthogonal to these ε-directions in the space of dual quaternions even after they acquire some ε-component via rotation or translation. Thus, what we see in projections into 2-dimensional Euclidean space is not an accurate depiction of the geometry of the 8-dimensional space of dual quaternions. Rather these pictures are an accurate depiction of how we construe these transformations when we interpret these transformations on points and vectors in 3-dimensions. Nevertheless, with these caveats in mind, these pictures are still quite instructive.

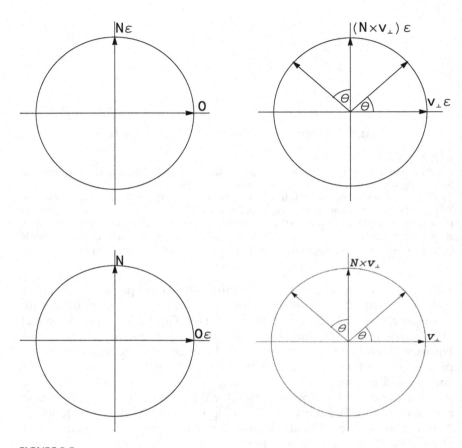

FIGURE 8.5

Rotations in the space of dual quaternions. Here N is a unit (normal) vector, $v_\perp$ is a (normal) vector perpendicular to N, and $q = q(N, \theta/2) = \cos(\theta/2)O + \sin(\theta/2)N$ is a unit quaternion. The map $x \mapsto qxq^\dagger$ leaves the $O(N\varepsilon)$ – plane and the $(O\varepsilon)N$ – plane unchanged and rotates vectors in the $v_\perp(N \times v_\perp)$ -plane and the $(v_\perp\varepsilon)(N \times v_\perp\varepsilon)$-plane by the angle θ. Thus, the map $x \mapsto qxq^\dagger$ represents a rotation in the 8-dimensional space of dual quaternions.

Hence, the appearance of Figure 8.6 notwithstanding, the map $x \mapsto TxT^\dagger$ is an isometry in the 8-dimensional space of dual quaternions, since multiplication by unit dual quaternions preserves both lengths and angles. In particular, $|O + dt\varepsilon| = |O|$ and $(O + dt\varepsilon) \cdot t\varepsilon = 0$, so in 8-dimensions $O + dt\varepsilon$ has unit length and is orthogonal to the $t\varepsilon$-axis. Similarly, $|t + dO\varepsilon| = |t|$ and $(t + dO\varepsilon) \cdot O\varepsilon = 0$, so in 8-dimensions $t + dO\varepsilon$ has unit length and is orthogonal to the $O\varepsilon$-axis. Thus, shears in these planes do not change the angles between the axes in 8-dimensions (see Exercise 2).

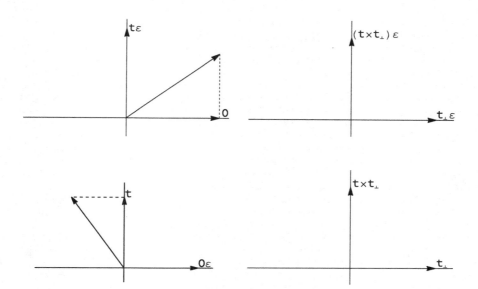

FIGURE 8.6
Translations as shears in the space of dual quaternions. Here $t\varepsilon$ is a unit vector, and the map $T = T(t, d/2) = O + \frac{1}{2} dt\varepsilon$ represents a translation in the direction $t\varepsilon$ by the distance $d > 0$. Thus, for vectors, the map $v\varepsilon \mapsto Tv\varepsilon T^\dagger = v\varepsilon$ is the identity. Moreover, for normal vectors, the map $v \mapsto TvT^\dagger = v - d(t \cdot v)O\varepsilon$ is the identity on all normal vectors v perpendicular to the vector $t\varepsilon$. Therefore the $t_\perp (t \times t_\perp)$-planes and the $(t_\perp \varepsilon)(t \times t_\perp \varepsilon)$-planes are unchanged. But the map $t \mapsto TtT^\dagger = t - dO\varepsilon$ appears to shear vectors in the $(O\varepsilon)t$-plane in the negative horizontal direction $(-d < 0)$, and the map $O \mapsto TOT^\dagger = O + dt\varepsilon$ appears to shear vectors in $O(t\varepsilon)$-plane in the positive vertical direction $(d > 0)$.

Exercises

1. Using the Taylor expansion for e^x, show that $e^{\varepsilon dt} = O + dt\varepsilon$.
2. Show that:
   a. $|O + dt\varepsilon| = |O|$
   b. $|t + dO\varepsilon| = |t|$
   c. $(O + dt\varepsilon) \cdot t\varepsilon = 0$
   d. $(t + dO\varepsilon) \cdot O\varepsilon = 0$
3. Draw the figures of the four mutually orthogonal planes in the space of dual quaternions for the map $x \mapsto qxq$, where $q = q(N, \theta/2) = \cos(\theta/2)O + \sin(\theta/2)N$.

1.9

Matrices versus Dual Quaternions

Matrices provide an alternative way to compute translations, rotations, screw transformations, and reflections as well as perspective and pseudo-perspective. Therefore, it is natural to compare representations and computations using homogeneous coordinates and 4×4 matrices to representations and computations using dual quaternions.

1.9.1 Representations and Computations with Matrices and Dual Quaternions

Here we compare and contrast the compactness of representations and the speed of computations using 4×4 matrices and dual quaternions.

Representations

- Matrices of order 4×4 are represented by 16 scalar entries.
- Dual quaternions are represented by a pair of quaternions, which require a total of 8 scalar values.

Thus, dual quaternion representations are more compact than 4×4 matrix representations for rigid motions and projective transformations.

Compositions

- Multiplying two 4×4 matrices representing rigid motions requires 48 scalar multiplications (see Exercise 1).
- Multiplying two dual quaternions representing rigid motions uses 3 quaternion products, each requiring 16 scalar multiplications for a total of $3 \times 16 = 48$ scalar multiplications.

Thus, composing two rigid motions by dual quaternion multiplication is computationally equivalent to composing two rigid motions by matrix multiplication.

DOI: 10.1201/9781003398141-10

Transformations

- Multiplying 4 homogeneous coordinates representing an affine point by a 4×4 matrix representing a rigid motion uses 12 scalar multiplications (see Exercise 2).
- Sandwiching a dual quaternion between two dual quaternions uses 2 dual quaternion products, each requiring 48 scalar multiplications for a total of $2 \times 48 = 96$ scalar multiplications.

Thus, computing rigid motions by dual quaternion multiplication is computationally much less efficient than computing rigid motions by matrix multiplication.

Clever implementations can reduce somewhat some of these computational costs, but these comparisons give a good rough idea of the tradeoffs. We conclude then that although dual quaternion representations for rigid motions are twice as compact as matrix representations, matrix representations are far more efficient for computing rigid motions and equally as efficient for composing these transformations.

The main advantage of dual quaternions is insight not computation. It is often easier to understand the results of manipulating dual quaternions than manipulating 4×4 matrices.

Nevertheless, dual quaternions also have some computational advantages. With dual quaternions we can use ScLERP to interpolate between two rigid motions; there is no such simple, effective method to interpolate between two rigid motions represented by 4×4 matrices. Also, we can renormalize dual quaternions that get distorted due to floating point arithmetic. There is no simple, effective way to renormalize rigid motions represented by 4×4 matrices.

Exercises

1. Explain why multiplying two 4×4 matrices representing rigid motions requires only 48 rather than 64 scalar multiplications.
2. Explain why multiplying 4 homogeneous coordinates representing a point in affine space by a 4×4 matrix representing a rigid motion requires only 12 rather than 16 scalar multiplications.
3. Redo the computation counts comparing dual quaternion and 4×4 matrix multiplication for perspective and pseudo-perspective.
4. Let p_1, q_1 be two quaternions. Since multiplication by a quaternion is a linear transformation in the space of quaternions, there are two 4×4 matrices $L(q_1), R(p_1)$ [9, Chapter 8] such that

$$p_1 q_1 = q_1 * L(p_1) = p_1 * R(q_1)$$

Let $p = p_1 + p_2\varepsilon$ and $q = q_1 + q_2\varepsilon$ be two dual quaternions. Show that:

a. $pq = (q_1, q_2)* \begin{pmatrix} L(p_1) & L(p_2) \\ 0 & L(p_1) \end{pmatrix}$

$= (p_1, p_2)* \begin{pmatrix} R(q_1) & R(q_2) \\ 0 & R(q_1) \end{pmatrix}.$

b. multiplying the dual quaternions p, q directly using dual quaternion multiplication or multiplying the dual quaternions p, q using matrix multiplication takes the same number of scalar multiplications.

1.9.2 Converting between Matrices and Dual Quaternions

Since dual quaternion representations are more compact than matrix representations and computations with matrices are often faster than computations with dual quaternions, we want to be able to convert between these two representations both for rigid motions and for perspective projections.

1.9.2.1 Rigid Motions

In order to perform these conversions for rigid motions, we begin by observing that every rigid motion is equivalent to a unit dual quaternion. Then we shall show how to convert between unit dual quaternions and matrix representations for rigid motions.

Theorem 9.1

Every rigid motion is equivalent to a unit dual quaternion.

Proof: By Lemma 4.6, every rigid motion is the product of a rotation about the origin followed by an arbitrary translation. But rotations about the origin are represented by unit quaternions, and translations are represented by unit dual quaternions. Since the length of the product of two dual quaternions is equal to the product of their lengths, it follows that every rigid motion can be represented by a unit dual quaternion.

Conversely, let $q = q_1 + q_2\varepsilon$ be a unit dual quaternion. Then we can factor q by setting

$$q = q_1 + q_2\varepsilon = \left(O + \frac{1}{2} 2 q_2 q_1^* \varepsilon \right) q_1 \tag{9.1}$$

To show that this factoring is correct, observe that since q is a unit dual quaternion $|q| = O$, so by Equation (2.4) $|q_1| = O$; therefore $q_1^* q_1 = |q_1|^2 = O$. Since q_1 is a unit quaternion, q_1 represents a rotation about a line passing through the origin. To show that $O + \frac{1}{2} 2 q_2 q_1^* \varepsilon$ represents a translation, we need to show that $q_2 q_1^* \varepsilon$ is a vector. Since $|q| = O$, by Equation (2.4) we also have $q_1 \cdot q_2 = 0$. Therefore, since q_1, q_2 are quaternions

$$q_2 q_1^* = (m_2 O + v_2)(m_1 O - v_1) = (m_2 m_1 + v_2 \cdot v_1)O - m_2 v_1 + m_1 v_2 - v_2 \times v_1$$

$$= (q_1 \cdot q_2)O - m_2 v_1 + m_1 v_2 - v_2 \times v_1$$

But $q_1 \cdot q_2 = 0$, so

$$q_2 q_1^* = -m_2 v_1 + m_1 v_2 - v_2 \times v_1,$$

which is indeed a vector – that is, a pure quaternion – in 3-dimensions. Since $q_2 q_1^*$ is a pure quaternion, the dual quaternion $\left(O + \frac{1}{2} 2 q_2 q_1^* \varepsilon \right)$ represents a translation in 3-dimensions. Thus Equation (9.1) shows that every unit quaternion q can be factored into a rotation about a line passing through the origin followed by a translation. Therefore, every unit dual quaternion represents a rigid motion. ◊

Now let S be a 4×4 matrix representing a rigid motion. Then by Lemma 4.6 S can be represented by a rotation around a line passing through the origin followed by a translation. Therefore, there is a 3×3 rotation matrix $R = (R_{i,j})$ and a translation vector $v_T = \left(v_T^1, v_T^2, v_T^3 \right)$ such that

$$S = \begin{pmatrix} R & 0 \\ v_T & 1 \end{pmatrix}.$$

Our goal is to solve the following two problems:

i. Given two quaternions q_1 and q_2 representing a unit dual quaternion $q = q_1 + q_2\varepsilon$, find the 3×3 rotation matrix R and the translation vector v_T representing the equivalent 4×4 rigid motion matrix $S = \begin{pmatrix} R & 0 \\ v_T & 1 \end{pmatrix}$.

ii. Given a 3×3 rotation matrix R and a translation vector v_T representing a 4×4 rigid motion matrix $S = \begin{pmatrix} R & 0 \\ v_T & 1 \end{pmatrix}$, find the two quaternions q_1 and q_2 representing the equivalent unit dual quaternion $q = q_1 + q_2\varepsilon$.

To solve the first problem, observe that both the unit quaternion q_1 and the 3×3 rotation matrix R represent rotations about a line passing through the origin. Therefore, R depends only on q_1. The formula for the entries of $R = (R_{ij})$ in terms of the components of the quaternion $q_1 = (\alpha, \beta, \gamma, \delta)$ is derived in [9, Chapter 8]:

$$R = \begin{pmatrix} \delta^2 + \alpha^2 - \beta^2 - \gamma^2 & 2(\alpha\beta + \gamma\delta) & 2(\alpha\gamma - \beta\delta) \\ 2(\alpha\beta - \gamma\delta) & \delta^2 - \alpha^2 + \beta^2 - \gamma^2 & 2(\beta\gamma + \alpha\delta) \\ 2(\alpha\gamma + \beta\delta) & 2(\beta\gamma - \alpha\delta) & \delta^2 - \alpha^2 - \beta^2 + \gamma^2 \end{pmatrix}. \quad (9.2)$$

Moreover, by Equation (9.1)

$$v_T = 2q_2 q_1^*. \quad (9.3)$$

Conversely, given a 3×3 rotation matrix $R = (R_{i,j})$, it is shown in [9, Chapter 8] how to find the components of the corresponding unit quaternion $q_1 = (\alpha, \beta, \gamma, \delta)$ in terms of the entries of R. These formulas are summarized in Table 9.1, where θ is the angle of rotation.

Once we know q_1, we can easily solve for q_2 using Equation (9.3):

$$v_T = 2q_2 q_1^* \Rightarrow q_2 = \frac{1}{2} v_T q_1. \quad (9.4)$$

TABLE 9.1

The Components of the Unit Quaternion $q_1 = (\alpha, \beta, \gamma, \delta)$ in Terms of the Entries of the Corresponding Rotation Matrix $R = (R_{i,j})$

$\theta \neq \pi$	$\theta = \pi$
$\delta = \dfrac{1}{2}\sqrt{R_{1,1} + R_{2,2} + R_{3,3} + 1}$	$\alpha = \dfrac{1}{2}\sqrt{R_{1,1} - R_{2,2} - R_{3,3} + 1}$
$\alpha = \dfrac{R_{2,3} - R_{3,2}}{2\sqrt{R_{1,1} + R_{2,2} + R_{3,3} + 1}}$	$\beta = \dfrac{R_{1,2} + R_{2,1}}{2\sqrt{R_{1,1} - R_{2,2} - R_{3,3} + 1}}$
$\beta = \dfrac{R_{3,1} - R_{1,3}}{2\sqrt{R_{1,1} + R_{2,2} + R_{3,3} + 1}}$	$\gamma = \dfrac{R_{1,3} + R_{3,1}}{2\sqrt{R_{1,1} - R_{2,2} - R_{3,3} + 1}}$
$\gamma = \dfrac{R_{1,2} - R_{2,1}}{2\sqrt{R_{1,1} + R_{2,2} + R_{3,3} + 1}}$	$\delta = \dfrac{R_{2,3} - R_{3,2}}{2\sqrt{R_{1,1} - R_{2,2} - R_{3,3} + 1}}$

Here θ is the angle of rotation.

Exercises

1. Show that the 4×4 matrix corresponding to the dual quaternion $T(N,d) = O + \dfrac{d}{2} N\varepsilon$ that represents translation by the distance d along the unit vector $N\varepsilon$ is

$$Trans(N,d) = \begin{pmatrix} I & 0 \\ dN & 1 \end{pmatrix},$$

   where I denotes the 3×3 identity matrix.

2. Let C be a point and let N be a unit normal vector. Set

$$\pi = N - (C \cdot N).$$

   By Corollary 3.2 π represents the plane passing through the point C with unit normal vector N, and by Proposition 4.18 the map $P \mapsto -\pi P^* \pi$ reflects the point P in the plane π. Consider the 4×4 matrix

$$M = \begin{pmatrix} I - 2N^T N & 0 \\ 2(C \cdot N)N & 1 \end{pmatrix},$$

   where I denotes the 3×3 identity matrix.

   a. Show that the map $P \mapsto (P,1)^* M$ reflects the point P in the plane π passing through the point C perpendicular to the unit normal vector N.

   b. Given the dual quaternion π, find the entries of the matrix M.

   c. Given the entries of the matrix M, find the dual quaternion π.

1.9.2.2 Perspective and Pseudo-Perspective

Converting between dual quaternions and matrix representations for perspective and pseudo-perspective is actually much easier than converting between dual quaternions and matrix representations for rigid motions.

To understand how this conversion works, recall that the unit dual quaternion representing both perspective and pseudo-perspective is

$$T\left(N, \frac{-1}{2d}\right) = O - \frac{1}{2d} N\varepsilon,$$

where N is the unit normal to the perspective plane and d is the distance from the eye to the perspective plane. To apply this dual quaternion to compute perspective and pseudo-perspective, we compose this transformation with duality.

Here we shall show that the 4×4 matrix

$$Persp(N,d) = \begin{pmatrix} I & N^T/d \\ 0 & 1 \end{pmatrix},$$

where I denotes the 3×3 identity matrix and N^T denotes the transpose of N—that is, the column vector with the same entries as the row vector N, represents the same transformation on points and vectors as sandwiching points and vectors with the dual quaternion $T\left(N, \frac{-1}{2d}\right)$ composed with duality.

To verify this claim, we need only check that these two transformations give the same results at the origin and on three basis vectors. Keep in mind here that the dual to the origin O is $O\varepsilon$, that $(0,0,0,1)$ represents the origin O in homogeneous coordinates, and that $(N,0)$ represents the vector $N\varepsilon$ in homogeneous coordinates.

1. At the origin

$$\left(T\left(N, \frac{-1}{2d}\right)(O^*)T\left(N, \frac{-1}{2d}\right)\right)^{\#} = \left(\left(O - \frac{1}{2d}N\varepsilon\right)(O\varepsilon)\left(O - \frac{1}{2d}N\varepsilon\right)\right)^{\#}$$

$$= (O\varepsilon)^{\#} = O$$

$$O * Persp(N,d) = (0,0,0,1) * \begin{pmatrix} I & \dfrac{N^T}{d} \\ 0 & 1 \end{pmatrix} = (0,0,0,1) = O$$

2. On the unit vector $N\varepsilon$

$$\left(T\left(N, \dfrac{-1}{2d}\right)(N\varepsilon)^* T\left(N, \dfrac{-1}{2d}\right) \right)^{\#} = \left(\left(O - \dfrac{1}{2d}N\varepsilon\right)N\left(O - \dfrac{1}{2d}N\varepsilon\right) \right)^{\#}$$

$$= \left(N + \dfrac{1}{d}O\varepsilon\right)^{\#} = N\varepsilon + \dfrac{1}{d}O$$

$$(N,0) * Persp(N,d) = (N,0) * \begin{pmatrix} I & N^T/d \\ 0 & 1 \end{pmatrix}$$

$$= \left(N, \dfrac{1}{d}(N \cdot N)\right) = (N,0) + \dfrac{1}{d}O$$

3. On vectors $v_\perp \varepsilon$ perpendicular to $N\varepsilon$

$$\left(T\left(N, \dfrac{-1}{2d}\right)(v_\perp \varepsilon)^* T\left(N, \dfrac{-1}{2d}\right) \right)^{\#} = \left(\left(O - \dfrac{1}{2d}N\varepsilon\right)v_\perp\left(O - \dfrac{1}{2d}N\varepsilon\right) \right)^{\#}$$

$$= \left(\left(v_\perp - \dfrac{1}{2d}Nv_\perp\varepsilon\right)\left(O - \dfrac{1}{2d}N\varepsilon\right) \right)^{\#}$$

$$= \left(v_\perp - \dfrac{1}{d}\left(\dfrac{Nv_\perp + v_\perp N}{2}\right)\varepsilon \right)^{\#}$$

$$= \left(v_\perp + \dfrac{1}{d}(N \cdot v_\perp)O\varepsilon \right)^{\#} = v_\perp \varepsilon$$

$$(v_\perp,0) * Persp(N,d) = (v_\perp,0) * \begin{pmatrix} I & N^T/d \\ 0 & 1 \end{pmatrix}$$

$$= \left(v_\perp, \dfrac{1}{d}(v_\perp \cdot N) \right) = (v_\perp,0)$$

Therefore, by linearity the dual quaternion $T\left(N,\dfrac{-1}{2d}\right)=O-\dfrac{1}{2d}N\varepsilon$ com-

posed with duality and the 4×4 matrix $Persp(N,d)=\begin{pmatrix} I & N^T/d \\ 0 & 1 \end{pmatrix}$

represent the same transformation.

Thus, we have the following simple correspondence:

$$T\left(N,\dfrac{-1}{2d}\right)=O-\dfrac{1}{2d}N\varepsilon \;\;\leftrightarrow\;\; Persp(N,d)=\begin{pmatrix} I & N^T/d \\ 0 & 1 \end{pmatrix}.$$

Converting between these two representations for perspective and pseudo-perspective is quite simple: we merely take the transpose of a vector and multiply or divide the components by a factor of -2.

Exercises

1. Show that in 4-dimensions the 4×4 matrices $\begin{pmatrix} I & N^T/d \\ 0 & 1 \end{pmatrix}$ and

   $\begin{pmatrix} I & 0 \\ dN & 1 \end{pmatrix}$ represent two different shears in the NO-plane.

2. Let R be a 3×3 matrix representing a rotation around a line passing through the origin. Show that the 4×4 matrix representing this rotation followed by perspective projection from the eye at the origin to a plane with unit normal N at a distance d from the eye is

$$RPersp(R,N,d)=\begin{pmatrix} R & R*\left(N^T/d\right) \\ 0 & 1 \end{pmatrix}.$$

3. In this exercise we explore the relationship between unit dual quaternions composed with duality and the corresponding 4×4 matrices.

   a. Let $q=q_1+q_2\varepsilon$ be a unit dual quaternion. Using Equation (9.1), show that the map $P\rightarrow\left(q(P-O)^{\#}q^{\dagger}\right)^{\#}$, represents a rotation corresponding to the unit quaternion q_1 followed by a perspective projection from the eye at the origin to the plane with unit normal $N=-\dfrac{q_2\overset{*}{q_1}}{|q_2|}$ at a distance $d=\dfrac{1}{2|q_2|}$ from the eye.

b. Let $RPersp = \begin{pmatrix} R & v^T \\ 0 & 1 \end{pmatrix}$ be the 4×4 matrix representing the

same transformation as the unit dual quaternion $q = q_1 + q_2\varepsilon$ composed with duality as in part a: a rotation R about a line through the origin followed by perspective projection with the eye at the origin. Using part a and Exercise 2, show that $-2q_2\overset{*}{q_1} = v * R$

c. Using part b, show how to convert between the dual quaternion and matrix representations for rotation followed by perspective projection.

1.10

Insights

Here we compare some properties of quaternions and dual quaternions and present as well some of the main insights we came across while investigating dual quaternions.

Comparisons

TABLE 10.1

Comparisons Between Quaternions and Dual Quaternions

	Quaternions	Dual Quaternions
Dimension	4-dimensions	8-dimensions
Coefficients	Real numbers	Dual real numbers
Zero divisors	None	$\varepsilon^2 = 0$
Points	$P = O + v_p$	$P = O + v_p\varepsilon$
Vectors	v	$v\varepsilon$
Planes	$\pi = dO + v_\pi$	$\pi = dO\varepsilon + v_\pi$
Transformations	Rotations	Rotations and translations
Interpolation	SLERP	ScLERP

Extensions

- Quaternions are extensions of the complex numbers to 4-dimensions.
- Dual quaternions are extensions of the dual real numbers to 8-dimensions.

Representations

- Quaternions do not distinguish clearly between mass-points and planes.
- Dual quaternions do distinguish clearly between mass-points and planes.

Isometries

- Unit quaternions are isometries in 4-dimensions.
- Unit dual quaternions are isometries in 8-dimensions.

Transformations

- Unit quaternions represent rotations in 3-dimensions.
- Unit dual quaternions represent rigid motions in 3-dimensions.
- Rigid motions in 3-dimensions are equivalent to screw transformations.

Shear

- Translations are shears in 4-dimensions. Shearing the origin in 4-dimensions translates points in 3-dimensions.
- Perspective projections are shears in 4-dimensions. Shearing vectors in 4-dimensions introduces mass at the origin.

Interpolation

- SLERP interpolates between two unit quaternions in order to interpolate between two rotations in 3-dimensions.
- ScLERP interpolates between two unit dual quaternions in order to interpolate between two screw transformations (rigid motions) in 3-dimensions.

Duality

- Converts distance to mass and mass into distance
- Changes translations to perspective projections

1.11

Formulas

We close with a summary for easy future reference of the main formulas we have encountered for dual quaternions.

1.11.1 Algebra

Conjugates

$$\text{Star}: (q_1 + q_2\varepsilon)^* = q_1^* + q_2^*\varepsilon$$

$$\text{Bar}: \overline{q_1 + q_2\varepsilon} = q_1 - q_2\varepsilon$$

$$\text{Cross}: q^\dagger = q_1^* - q_2^*\varepsilon$$

Conjugates and Products

$$\text{Star}: (pq)^* = q^* p^*$$

$$\text{Bar}: \overline{pq} = \overline{p}\,\overline{q}$$

$$\text{Cross}: (pq)^\dagger = q^\dagger p^\dagger$$

Dot Product, Cross Product, and Triple Product *for pure quaternions u, v representing vectors in 3-dimensions*

$$(u \cdot v)O = -\left(\frac{uv + vu}{2}\right)$$

$$u \times v = \frac{uv - vu}{2}$$

$$vuv = (v \cdot v)u - 2(v \cdot u)v$$

DOI: 10.1201/9781003398141-12

Dot Products *of the dual quaternions* $p = p_1 + p_2\varepsilon$, $q = q_1 + q_2\varepsilon$, $r = r_1 + r_2\varepsilon$

$$p \cdot q = \frac{pq^* + qp^*}{2}$$

$$p \cdot q = (p_1 \cdot q_1)O + (p_1 \cdot q_2 + p_2 \cdot q_1)\varepsilon$$

$$p \cdot (q + r) = p \cdot q + p \cdot r$$

$$(p + q) \cdot r = p \cdot r + q \cdot r$$

$$(pq) \cdot (pr) = |p|^2 (q \cdot r)$$

$$p \perp q \Leftrightarrow p_1 \cdot q_1 = 0 \Leftrightarrow p_1 \perp q_1$$

Lengths *of the dual quaternions* $p = p_1 + p_2\varepsilon$, $q = q_1 + q_2\varepsilon$. *Here* $d = a + b\varepsilon$ *is a dual real number.*

$$|q|^2 = q \cdot q = qq^* = |q_1|^2 O + 2(q_1 \cdot q_2)\varepsilon$$

$$|q| = |q_1|O + \frac{q_1 \cdot q_2}{|q_1|}\varepsilon$$

$$|pq| = |p||q|$$

$$|dq|^2 = d^2 |q|^2$$

Angle *between the dual quaternions* $q = q_1 + q_2\varepsilon$ *and* $r = r_1 + r_2\varepsilon$

$$\cos(\theta_{q,r}) = \frac{q \cdot r}{|q||r|} \text{ provided that } |q||r| \neq 0$$

$$\cos(\theta_{q,r}) = 0 \Leftrightarrow q \perp r \Leftrightarrow q_1 \cdot r_1 = 0$$

$$\cos(\theta_{pq,pr}) = \frac{(pq) \cdot (pr)}{|pq||pr|} = \frac{q \cdot r}{|q||r|} = \cos(\theta_{q,r}) \text{ for } |p||q||r| \neq 0$$

$$\cos(\theta_{pq,pr}) = \cos(\theta_{q,r}) \text{ for } |p||q||r| \neq 0$$

Renormalization *of the dual quaternion* $q = q_1 + q_2\varepsilon$

$$\frac{q}{|q|} = \frac{|q_1|^2 q_1 O + \left(|q_1|^2 q_2 - (q_1 \cdot q_2)q_1\right)\varepsilon}{|q_1|^3}$$

1.11.2 Geometry

Representations

	Quaternions	Dual Quaternions
Points	$P = O + p_1 i + p_2 j + p_3 k$ $P \leftrightarrow O + v_P$	$P = O + (p_1 i + p_2 j + p_3 k)\varepsilon$ $P \leftrightarrow O + v_P \varepsilon$
Vectors	$v = v_1 i + v_2 j + v_3 k$ $v \leftrightarrow v$	$v = (v_1 i + v_2 j + v_3 k)\varepsilon$ $v \leftrightarrow v\varepsilon$
Planes	$\pi \equiv ax + by + cz + d = 0$ $\pi \leftrightarrow ai + bj + ck + dO$ $\pi \leftrightarrow dO + v_\pi$	$\pi \equiv ax + by + cz + d = 0$ $\pi \leftrightarrow ai + bj + ck + dO\varepsilon$ $\pi \leftrightarrow dO\varepsilon + v_\pi$
Lines	2 points or 2 planes	$l = v + u\varepsilon, \; v \cdot u = 0$

Plucker Coordinates and Dual Plucker Coordinates

$$l = v + (P - O) \times v$$

(v = vector parallel to line l and P = point on line l)

$$l = v_1 \times v_2 + (d_2 v_1 - d_1 v_2)\varepsilon$$

(l = intersection line of the two planes $\pi_1 = v_1 + d_1 O\varepsilon$ and $\pi_2 = v_2 + d_2 O\varepsilon$)

Incidence Relations

$$P \cdot \pi = 0$$

(point P on plane π)

$$(P - O) \times v = u\varepsilon$$

(point P on line $l = v + u\varepsilon$)

$$v \cdot v_\pi = 0 \text{ and } d = \frac{\det(u \; v \; v_\pi)}{|v|^2}$$

(line $l = v + u\varepsilon$ on plane $\pi = dO\varepsilon + v_\pi$)

Distance Formulas

$$dist(P, \pi) = (P \cdot \pi)^{\#}$$ *(from point P to plane π with unit normal vector)*

$$dist(P, l) = \left| \left(\frac{(P - O) \times v - u\varepsilon}{|v|} \right)^{\#} \right|$$ *(from point P to line $l = v + u\varepsilon$)*

Intersection Formulas

$$l = (\pi_2 - d_2 O\varepsilon) \times (\pi_1 - d_1 O\varepsilon) + (d_2\pi_1 - d_1\pi_2)\varepsilon = v_2 \times v_1 + (d_2 v_1 - d_1 v_2)\varepsilon$$

(*intersection line l of two planes* $\pi_1 = v_1 + d_1 O\varepsilon$ *and* $\pi_2 = v_2 + d_2 O\varepsilon$)

$$P = O + \frac{(v_1 \cdot C_1)(v_2 \times v_3) + (v_2 \cdot C_2)(v_3 \times v_1) + (v_3 \cdot C_3)(v_1 \times v_2)}{\det(v_1 \; v_2 \; v_3)}$$

(*intersection point P of the three planes* $\pi_k = v_k - C_k \cdot v_k$, $k = 1, 2, 3$)

$$P = O + \frac{v \times u}{|v|^2}\varepsilon - \frac{d|v|^2 + \det(v \; u \; v_\pi)}{(v \cdot v_\pi)|v|^2}v\varepsilon, \text{ provided that } v \cdot v_\pi \neq 0$$

(*intersection point P of line* $l = v + u\varepsilon$ *with plane* $\pi = dO\varepsilon + v_\pi$)

$$P = O + w_2\varepsilon + \left(\frac{((w_2 - w_1) \times v_2) \cdot (v_2 \times v_1)}{|v_2 \times v_1|^2}v_2\right)\varepsilon$$

$$w_1 = \frac{v_1 \times u_1}{|v_1|^2}, w_2 = \frac{v_2 \times u_2}{|v_2|^2}$$

(*intersection point P of two intersecting lines* $l_1 = v_1 + u_1\varepsilon$ *and* $l_2 = v_2 + u_2\varepsilon$)

1.11.3 Duality

Basis Elements

$i^{\#} = i\varepsilon$	$(i\varepsilon)^{\#} = i$
$j^{\#} = j\varepsilon$	$(j\varepsilon)^{\#} = j$
$k^{\#} = k\varepsilon$	$(k\varepsilon)^{\#} = k$
$O^{\#} = O\varepsilon$	$(O\varepsilon)^{\#} = O$

Geometric Elements

Points	$P = O + (p_1 i + p_2 j + p_3 k)\varepsilon$	$P^{\#} = O\varepsilon + (p_1 i + p_2 j + p_3 k)$
	$P \leftrightarrow O + v_P \varepsilon$	$P^{\#} \leftrightarrow O\varepsilon + v_P$
Vectors	$v = (v_1 i + v_2 j + v_3 k)\varepsilon$	$v^{\#} = v_1 i + v_2 j + v_3 k$
	$v \leftrightarrow v\varepsilon$	$v^{\#} \leftrightarrow v$
Planes	$\pi = ai + bj + ck + dO\varepsilon$	$\pi^{\#} = (ai + bj + ck)\varepsilon + dO$
	$\pi \leftrightarrow v_\pi + dO\varepsilon$	$\pi^{\#} \leftrightarrow v_\pi \varepsilon + dO$
Lines	$l = (v_1 i + v_2 j + v_3 k)$	$l^{\#} = (v_1 i + v_2 j + v_3 k)\varepsilon$
	$\quad + (u_1 i + u_2 j + u_3 k)\varepsilon$	$\quad + (u_1 i + u_2 j + u_3 k)$
	$l = v + u\varepsilon$	$l^{\#} \leftrightarrow u + v\varepsilon$

Arbitrary Dual Quaternions

$$\left(q_1 + q_2 \varepsilon\right)^{\#} = q_2 + q_1 \varepsilon$$

$$(q^{\#})^{\#} = q$$

1.11.4 Transformations

Translation *by the distance d in the direction of the unit normal vector t*

$$T = T\left(t, \frac{d}{2}\right) = O + \frac{d}{2} t\varepsilon$$

$$P \mapsto TPT^{\dagger} \quad (\textit{points and planes})$$

$$l \mapsto TlT^{-1} \quad (\textit{lines})$$

Rotation *by the angle θ about the line L through the origin O parallel to the direction of the unit normal vector N*

$$q = q(N, \theta/2) = \cos(\theta/2)O + \sin(\theta/2)N$$

$$P \mapsto qPq^{\dagger} \quad (\textit{points, vectors, and planes})$$

$$l \mapsto qlq^{-1} \quad (\textit{lines})$$

Rotation *by the angle θ about the line L through the point C parallel to the direction of the unit normal vector N*

$$R = R(N,\theta,C) = \cos(\theta/2)O + \sin(\theta/2)\big(N + (C-O)\times N\big)\varepsilon$$

$$P \mapsto RPR^{\dagger} \quad (\textit{points, vectors, and planes})$$

$$l \mapsto RlR^{-1} \quad (\textit{lines})$$

Screw Transformation *rotation by the angle θ about the line L through the point C parallel to the direction of the unit normal vector N followed by translation by the distance d in the direction of the unit normal vector N*

$$S = S(N,\theta,C,d) = R(N,\theta,C) + \frac{1}{2}d\big(\cos(\theta/2)N - \sin(\theta/2)O\big)\varepsilon$$

$$P \mapsto SPS^{\dagger} \quad (\textit{points and planes})$$

$$l \mapsto SlS^{-1} \quad (\textit{lines})$$

Reflection *in a plane π with unit normal*

$$P \mapsto -\pi P^{*}\pi \quad (\textit{points, vectors, and planes})$$

$$l \mapsto \pi l^{\dagger}\pi^{-1} \quad (\textit{lines})$$

Perspective *from the eye E at the origin O to the plane π at a distance d from the eye perpendicular to the unit normal vector N*

$$T = T\left(N,-\frac{1}{2d}\right) = O - \frac{1}{2d}N\varepsilon$$

$$P \mapsto \big(T(P-E)^{\#}T\big)^{\#}$$

Pseudo-Perspective *with the eye at $E = O - dN\varepsilon$ looking in the direction of the unit vector Nε*

$$T = T\left(N,-\frac{1}{2d}\right) = O - \frac{1}{2d}N\varepsilon$$

$$P \mapsto \big(TP^{\#}T\big)^{\#}$$

1.11.5 Interpolation

ScLERP *where S_0, S_1 are two dual quaternions, each representing a screw transformation*

$$S_1 S_0^* = \cos(\theta)O + \sin(\theta)N = \cos(\theta_1 + \theta_2\varepsilon)O + \sin(\theta_1 + \theta_2\varepsilon)(N_1 + N_2\varepsilon)$$

$$ScLERP(S_0, S_1, t) = \cos(t\theta)S_0 + \sin(t\theta)NS_0$$

Trigonometric Identities *for dual real angles*

$$\cos(\theta_1 + \theta_2\varepsilon) = \cos(\theta_1) - \theta_2\sin(\theta_1)\varepsilon$$

$$\sin(\theta_1 + \theta_2\varepsilon) = \sin(\theta_1) + \theta_2\cos(\theta_1)\varepsilon$$

1.11.6 Conversion Formulas

Rigid Motions

   i. From a unit dual quaternion $q = q_1 + q_2\varepsilon$ to a 4×4 rigid motion matrix

$$S = \begin{pmatrix} R & 0 \\ v_T & 1 \end{pmatrix}, \text{ where } q_1 = (\alpha, \beta, \gamma, \delta)$$

$$R = \begin{pmatrix} \delta^2 + \alpha^2 - \beta^2 - \gamma^2 & 2(\alpha\beta + \gamma\delta) & 2(\alpha\gamma - \beta\delta) \\ 2(\alpha\beta - \gamma\delta) & \delta^2 - \alpha^2 + \beta^2 - \gamma^2 & 2(\beta\gamma + \alpha\delta) \\ 2(\alpha\gamma + \beta\delta) & 2(\beta\gamma - \alpha\delta) & \delta^2 - \alpha^2 - \beta^2 + \gamma^2 \end{pmatrix}$$

$$v_T = 2q_2 q_1^*$$

   ii. From a 4×4 rigid motion matrix $S = \begin{pmatrix} R & 0 \\ v_T & 1 \end{pmatrix}$ to a unit dual quaternion $q = q_1 + q_2\varepsilon$, where $q_1 = (\alpha, \beta, \gamma, \delta)$, and θ is the angle of rotation.

$\theta \neq \pi$	$\theta = \pi$
$\delta = \dfrac{1}{2}\sqrt{R_{1,1} + R_{2,2} + R_{3,3} + 1}$	$\alpha = \dfrac{1}{2}\sqrt{R_{1,1} - R_{2,2} - R_{3,3} + 1}$
$\alpha = \dfrac{R_{2,3} - R_{3,2}}{2\sqrt{R_{1,1} + R_{2,2} + R_{3,3} + 1}}$	$\beta = \dfrac{R_{1,2} + R_{2,1}}{2\sqrt{R_{1,1} - R_{2,2} - R_{3,3} + 1}}$
$\beta = \dfrac{R_{3,1} - R_{1,3}}{2\sqrt{R_{1,1} + R_{2,2} + R_{3,3} + 1}}$	$\gamma = \dfrac{R_{1,3} + R_{3,1}}{2\sqrt{R_{1,1} - R_{2,2} - R_{3,3} + 1}}$
$\gamma = \dfrac{R_{1,2} - R_{2,1}}{2\sqrt{R_{1,1} + R_{2,2} + R_{3,3} + 1}}$	$\delta = \dfrac{R_{2,3} - R_{3,2}}{2\sqrt{R_{1,1} - R_{2,2} - R_{3,3} - 1}}$
$q_2 = \dfrac{1}{2}v_T q_1$	

Translations

Between a dual quaternion representing translation by a distance d in the direction of a unit normal vector t and a 4×4 matrix representing the same translation.

$$T(t, d/2) = O + \frac{d}{2} t \varepsilon \quad \leftrightarrow \quad Trans(t, d) = \begin{pmatrix} I & 0 \\ dt & 1 \end{pmatrix},$$

where I is the 3×3 identity matrix.

Perspective and Pseudo-Perspective Projections

Between a dual quaternion representing perspective or pseudo-perspective from a canonical eye point to a plane at a distance d from the eye with a unit normal N and a 4×4 matrix representing the same perspective or pseudo-perspective transformation.

$$T\left(N, -\frac{1}{2d}\right) = O - \frac{1}{2d} N \varepsilon \quad \leftrightarrow \quad Persp(N, d) = \begin{pmatrix} I & N^T/d \\ 0 & 1 \end{pmatrix},$$

where I is the 3×3 identity matrix, and N^T denotes the transpose of N.

Appendix: Cross Products

Notation

- u, v are vectors in 3-dimensions
- θ = angle from u to v

Definition

- $u \times v \perp u, v$
- $|u \times v| = |u||v| \, |\sin(\theta)|$
- $u, v, u \times v$ satisfy the right-hand rule

Basis Vectors

- $i \times j = k, \quad j \times k = i, \quad k \times i = j$

Anticommutative

- $u \times v = -v \times u$
- $u \times u = 0$

Nonassociative

- $u \times (v \times w) = (u \cdot w)v - (u \cdot v)w$
- $(u \times v) \times w = (w \cdot u)v - (w \cdot v)u$

Distributive

- $u \times (v + w) = u \times v + u \times w$
- $(v + w) \times u = v \times u + w \times u$

Jacobi Identity

- $u \times (v \times w) + v \times (w \times u) + w \times (u \times v) = 0$

Triple Product

- $u \cdot (v \times w) = \det(u \ \ v \ \ w)$
- $u \cdot (u \times v) = v \cdot (u \times v) = 0$

Orthogonality Conditions

- $u \cdot v = 0 \Rightarrow |u \times v| = |u||v|$
- $u \cdot v = 0 \Rightarrow \dfrac{v}{|v|} \times \left(\dfrac{u}{|u|} \times \dfrac{v}{|v|} \right) = \dfrac{u}{|u|}$

Conjugation

- $q(u \times v)q^{-1} = \left(quq^{-1}\right) \times \left(qvq^{-1}\right)$ {q = any invertible linear transformation}

Part II

Clifford Algebras for Dual Quaternions

1

A Brief Review of Clifford Algebra

1 Goals of Clifford Algebra

Quaternions and dual quaternions are special cases of Clifford algebra. Both quaternions and dual quaternions allow us to multiply vectors in 3-dimensions, but these products lie in a vector space of 4-dimensions. Clifford algebras allow us to multiply vectors in n-dimensions, but these products lie in a vector space with 2^n-dimensions.

Clifford algebra is also a richer algebra, which allows for the direct representation of vectors, points, lines, and planes. Lengths and distances, meets and joins, incidence relations and intersections, orientation and duality are also readily expressed in Clifford algebra.

One of the goals of Clifford algebra is to clarify matters by separating operators from operands. The operands are geometry: vectors, points, lines, and planes. The operators are transformations: translations, rotations, reflections, and perspective projections. Clifford algebra uses the greater number of dimensions to distinguish between operators and operands.

We shall assume throughout that the reader is already familiar with the general framework of Clifford algebra. Nevertheless, to refresh the reader's memory and to fix our terminology and notation, we provide here a brief review of Clifford algebra. For further details, see [6,9,13].

2 A Brief Introduction to Clifford Algebra

Clifford algebra associates to every real n-dimensional vector space R^n a real vector space of dimension 2^n. To construct a Clifford algebra for R^n, let $e_1, \ldots, e_n$ be an orthonormal basis for R^n. We introduce 2^n canonical generators (basis multivectors) of a Clifford algebra for R^n by taking all the formal products $e_{i_1} \cdots e_{i_j}$, $i_1 < i_2 < \cdots < i_j$, as basis vectors. These products are called *blades*. The number of factors j is called the *grade* of the blade $e_{i_1} \cdots e_{i_j}$.

There are $\binom{n}{j}$ blades of grade j, so counting the scalar 1, a Clifford

DOI: 10.1201/9781003398141-14

algebra for R^n has 2^n basis vectors. The blades of grade j form a vector space of dimension $\binom{n}{j}$. In particular, the multivectors of dimensions 0 and n are vector spaces of dimension 1. The multivectors of dimension 0—constant multiples of 1—are called *scalars*; the multivectors of dimension n—constant multiples of $e_1 \cdots e_n$—are called *pseudoscalars*.

To extend the definition of multiplication to products of the basis vectors in any order and to capture orientation, we set

$$e_j e_i = -e_i e_j, \quad i \neq j. \tag{2.1}$$

To complete the definition of multiplication, we need to define the products e_i^2 for $i = 1, \dots, n$. These products are defined by setting

$$e_i^2 = c_i, \tag{2.2}$$

where each c_i is a scalar value. With these definitions, we can now add and multiply multivectors. Multiplication is associative and distributes through addition, but multiplication is not commutative.

Notice that the value of c_i may differ for different indices i. Typically $c_i = 0, 1, -1$ but other choices are possible (see Chapter 3, Section 1). Different choices for the scalars c_i, $i = 1, \dots, n$, lead to different Clifford algebras for R^n, corresponding to different geometries on the space R^n. These differences account, in part, for the very richness of Clifford algebra as a paradigm for representing diverse geometries.

Example 2.1: Complex Numbers, Dual Real Numbers, and Hyperbolic Numbers. *The simplest Clifford algebras are the Clifford algebras R^1. Let e_1 be a unit vector in R^1. Then there are three canonical cases: $e_1^2 = -1$ generates the complex numbers, $e_1^2 = 0$ gives the dual real numbers, and $e_1^2 = +1$ yields the hyperbolic numbers (see Part I, Section 1.1, Exercise 9).*

Example 2.2: Quaternions. *Quaternions are isomorphic to a Clifford algebra for R^2. Let e_1, e_2 be an orthonormal basis for R^2, and set $e_1^2 = e_2^2 = -1$. Then it is easy to verify that $(e_1 e_2)^2 = -1$, and the Clifford algebra generated by $1, e_1, e_2, e_1 e_2$ is isomorphic to the quaternion algebra generated by O, i, j, k.*

Example 2.3: Dual Quaternions. *Dual quaternions are isomorphic to a Clifford algebra for R^3. Let e_0, e_1, e_2 be an orthonormal basis for R^3, and set $e_0^2 = 0$, $e_1^2 = e_2^2 = -1$. Then once again the subalgebra generated by $1, e_1, e_2, e_1 e_2$ is isomorphic to the quaternion algebra generated by O, i, j, k. In addition, the subalgebra generated by $e_0, e_1 e_0, e_2 e_0, e_1 e_2 e_0$ is isomorphic to the rest of the dual quaternion algebra generated by $O\varepsilon, i\varepsilon, j\varepsilon, k\varepsilon$. Moreover, the multiplication between these two subalgebras of this Clifford algebra exactly mimics the multiplication between the corresponding subalgebras in the space of dual quaternions.*

One of the goals of Clifford algebra is to clarify matters by separating operators and operands. However, neither in Example 2.2 nor in Example 2.3 does this separation take place. Rather these models are simply identical to the standard approaches to quaternions and dual quaternions; the names of the entities are changed, but the algebras are identical. In [9, Chapters 14–18] it is shown how to separate operators and operands by embedding the quaternions in an 8-dimensional Clifford algebra for R^3.

Example 2.4: Quaternions Revisited. *Consider the following Clifford algebra for R^3. Let e_1, e_2, e_3 be an orthonormal basis for R^3, and set $e_1^2 = e_2^2 = e_3^2 = 1$. Then the subalgebra generated by $1, e_3 e_2, e_1 e_3, e_2 e_1$ is isomorphic to the quaternion algebra generated by O, i, j, k. The vectors e_1, e_2, e_3 represent unit vectors along the x, y, z coordinate axes, and the product $e_1 e_2 e_3$ can be invoked to represent the point at the origin. Sandwiching between unit quaternions and their conjugates can be used to compute rotations on points and vectors in 3-dimensions. Thus the even-grade elements, the quaternions, are the operators, and the odd-grade elements, points and vectors, are the operands. For further details, see [9, Chapters 14–18].*

In the remainder of this monograph, we will show how to embed the dual quaternions as subalgebras in 16-dimensional Clifford algebras for R^4, where we can once again distinguish clearly between operators and operands. In addition, the algebra of quaternions in Example 2.4 is a subalgebra of the dual quaternions inside these Clifford algebras. There are two ways to accomplish this embedding: the plane model and the point model. We shall explore each of these models in turn: the plane model in Chapter 2 and the point model in Chapter 3.

Exercise

1. Let e_1, e_2 be an orthonormal basis for R^2. Set $e_1^2 = e_2^2 = 1$ and $I = e_1 e_2$.
   a. Show that the subalgebra generated by $1, I$ is isomorphic to the complex numbers.
   b. Let $v = a_1 e_1 + a_2 e_2$ be a vector in R^2 and let $z = \cos\theta + I\sin\theta$ correspond to a complex number. Show that map $z \mapsto zv$ rotates the vector v by the angle θ in the xy-plane.

---

3 Basic Products: Clifford Product, Inner Product, and Outer Product

In any Clifford algebra for R^n there are three important products for the elements of grade 1: the *Clifford product*, the *inner product* (*dot product*), and the *outer product* (*wedge product*). To express these products explicitly, let $u_i, v_i, i = 1, \ldots, n$ be scalars and set

$$u = u_1 e_1 + \cdots + u_n e_n$$

$$v = v_1 e_1 + \cdots + b_n e_n$$

Clifford Product

$$uv = \left(u_1 e_1 + \cdots + u_n e_n \right)\left(v_1 e_1 + \cdots + b_n e_n \right) = \sum_{i=1}^{n} c_i u_i v_i + \sum_{i<j} \left(u_i v_j - u_j v_i \right) e_i e_j$$

Inner (Dot) Product

$$u \cdot v = \sum_{i=1}^{n} c_i u_i v_i$$

Outer (Wedge) Product

$$u \wedge v = \sum_{i<j} \left(u_i v_j - u_j v_i \right) e_i e_j$$

By construction, $u \cdot v$ has grade 0, $u \wedge v$ has grade 2, and uv has mixed grade. Moreover

$$uv = u \cdot v + u \wedge v \qquad (3.1)$$

$$u \cdot v = \frac{uv + vu}{2} \qquad (3.2)$$

$$u \wedge v = \frac{uv - vu}{2} \qquad (3.3)$$

Hence

$$u \cdot v = v \cdot u \qquad (3.4)$$

$$u \wedge v = -v \wedge u \qquad (3.5)$$

Now it is straightforward to show (see Exercise 1) that for any invertible element s

$$\left(sus^{-1}\right)\cdot\left(svs^{-1}\right) = s(u\cdot v)s^{-1} = u\cdot v \tag{3.6}$$

$$\left(sus^{-1}\right)\wedge\left(svs^{-1}\right) = s(u\wedge v)s^{-1} \tag{3.7}$$

Thus conjugation with an invertible element s is both an inner and an outer automorphism.

These seven formulas are pretty much all the algebra we shall need in our study of the Clifford algebras for dual quaternions. Notice, in particular, the similarities and differences between:

a. the wedge product and the cross product of two vectors;
b. the Clifford product and the quaternion product of two vectors;
c. the dot product and the wedge product in terms of the Clifford product, and the dot product and the cross product in terms of the quaternion product.

We shall have more to say about these similarities and differences in subsequent sections.

Exercise

1. Let u, v be elements of grade 1, and let s be an invertible element. Show that
   a. $\left(sus^{-1}\right)\cdot\left(svs^{-1}\right) = s(u\cdot v)s^{-1} = u\cdot v$
   b. $\left(sus^{-1}\right)\wedge\left(svs^{-1}\right) = s(u\wedge v)s^{-1}$

(Hint: Express the dot product and the wedge product in terms of the Clifford product.)

3.1 Exterior Algebra: The Outer (Wedge) Product for Arbitrary Grades

The outer (wedge) product that we encountered in the previous section for two elements of grade 1 can be extended to two elements of arbitrary grade. Succinctly, if U has grade r and V has grade s, then by definition

$$U \wedge V = \text{the element of grade } r+s \text{ in the Clifford product } UV. \tag{3.8}$$

Notice that this definition is consistent with our definition of the wedge product for two elements of grade 1.

This extension of the wedge product inherits several of the properties of the Clifford product. For example, the wedge product is associative and distributes through addition. Moreover by Equation (3.5), for elements of grade 1,

$$e_j \wedge e_i = -e_i \wedge e_j, \text{ for all } i, j. \tag{3.9}$$

Notice how this anticommutativity formula for the wedge product differs somewhat from the corresponding formula for the Clifford product, where by Equation (2.1) this anticommutativity property holds only when $i \neq j$ because $e_i \cdot e_j = 0$. As a consequence of Equation (3.9)

$$e_i \wedge e_i = 0 \text{ for all } i. \tag{3.10}$$

In contrast, the Clifford product e_i^2 can be any nonzero scalar. The algebra generated by the wedge product is called the *exterior algebra* on R^n.

The following two propositions clarify the relationship between the Clifford product and the wedge product for products consisting of two or more elements of an orthonormal basis.

Proposition 3.1

Let $e_1, \ldots, e_n$ be an orthonormal basis for R^n. Then

$$e_{i_1} \wedge \cdots \wedge e_{i_r} = e_{i_1} \cdots e_{i_r} \quad \text{when } i_1 \neq i_2 \neq \cdots \neq i_r. \tag{3.11}$$

Proof: This result follows by induction on r. When $r = 2$, by assumption $e_1 \cdot e_2 = 0$. Therefore by Equation (3.1), $e_1 \wedge e_2 = e_1 e_2$. Suppose then that

$$e_{i_1} \wedge \cdots \wedge e_{i_r} = e_{i_1} \cdots e_{i_r} \quad \text{when } i_1 \neq i_2 \neq \cdots \neq i_r$$

and suppose too that $i_{r+1} \neq i_1, i_2, \ldots, i_r$. Then by Equation (3.8) and the inductive hypothesis

$$(e_{i_1} \wedge \cdots \wedge e_{i_r}) \wedge e_{i_{r+1}}$$

$$= \text{element of grade } r + 1 \text{ in the Clifford product } \left(e_{i_1} \wedge \cdots \wedge e_{i_r} \right) e_{i_{r+1}}$$

$$= \text{element of grade } r + 1 \text{ in the Clifford product } \left(e_{i_1} \cdots e_{i_r} \right) e_{i_{r+1}}$$

But $e_1, e_2, \ldots, e_n$ is an orthonormal basis, so $e_{i_1} e_{i_2} \cdots e_{i_{r+1}}$ is a blade of grade $r+1$. ◊

Proposition 3.2

Let $e_1, \ldots, e_n$ be an orthonormal basis for R^n. Then

$e_{i_1} \wedge \cdots \wedge e_{i_r} = 0$ whenever there is at least one pair of common factors. (3.12)

Proof: Again this result follows by induction on r. When $r = 2$, by Equation (3.5)

$$e_1 \wedge e_1 = -e_1 \wedge e_1 \Rightarrow e_1 \wedge e_1 = 0.$$

Now suppose that this result is true for r and consider $e_{i_1} \wedge \cdots \wedge e_{i_r} \wedge e_{i_{r+1}}$. There are 2 cases.

   Case 1: There is at least one pair of common factors among $e_{i_1}, \ldots, e_{i_r}$. Then by the inductive hypothesis

$$e_{i_1} \wedge \cdots \wedge e_{i_r} \wedge e_{i_{r+1}} = \left(e_{i_1} \wedge \cdots \wedge e_{i_r} \right) \wedge e_{i_{r+1}} = 0.$$

   Case 2: There is no pair of common factors among $e_{i_1}, \cdots, e_{i_r}$, but $i_{r+1} = i_j$ for some $j \leq r$. Then by Proposition 3.1

$$e_{i_1} \wedge \cdots \wedge e_{i_r} = e_{i_1} \cdots e_{i_r}$$

and by Equation (3.8)

$(e_{i_1} \wedge \cdots \wedge e_{i_r}) \wedge e_{i_{r+1}}$

   $=$ element of grade $r + 1$ in the Clifford product $\left(e_{i_1} \wedge \cdots \wedge e_{i_r} \right) e_{i_{r+1}}$

But by Equations (2.1) and (2.2)

$$\left(e_{i_1} \wedge \cdots \wedge e_{i_r} \right) e_{i_{r+1}} = \left(e_{i_1} \cdots e_{i_r} \right) e_{i_{r+1}} = \pm e_{i_j}^2 \left(e_{i_1} \cdots e_{i_{j-1}} e_{i_{j+1}} e_{i_r} \right),$$

and $e_{i_j}^2$ is a scalar, so $\pm e_{i_j}^2 \left(e_{i_1} \cdots e_{i_{j-1}} e_{i_{j+1}} e_{i_r} \right)$ has grade r. Therefore, there is no element of grade $r + 1$ in the Clifford product $\left(e_{i_1} \wedge \cdots \wedge e_{i_r} \right) e_{i_{r+1}}$, so $(e_{i_1} \wedge \cdots \wedge e_{i_r}) \wedge e_{i_{r+1}} = 0$. ◊

The next proposition presents yet another useful relation between the wedge product and the Clifford product.

Proposition 3.3

Let U be an element of grade r and let V be an element of grade s. Then

$$U \wedge V = \frac{UV + (-1)^{rs} VU}{2} \tag{3.13}$$

Proof: We will prove this result for the blades

$$U = e_{i_1} \cdots e_{i_r}$$

$$V = e_{j_1} \cdots e_{j_s}.$$

The general result then follows by linearity, since the wedge product distributes through addition. Now we consider two cases.

Case 1: U and V have no common factors. In this case, counting sign flips, we find that

$$\left(e_{i_1} \cdots e_{i_r}\right)\left(e_{j_1} \cdots e_{j_s}\right) = (-1)^{rs}\left(e_{j_1} \cdots e_{j_s}\right)\left(e_{i_1} \cdots e_{i_r}\right)$$

$$UV = (-1)^{rs} VU.$$

Therefore since by definition

$$U \wedge V = \text{the element of grade } r + s \text{ in the Clifford product } UV$$

$$U \wedge V = \frac{UV + (-1)^{rs} VU}{2}.$$

Case 2: U and V have a common factor. In this case by Proposition 3.2, $U \wedge V = 0$, so we must show that the right-hand side of Equation (3.13) is also zero. Without loss of generality, we can assume that $e_{i_1} = e_{j_1}$. Let $e_{i_1} e_{j_1} = c$. Now counting sign flips, we find that:

$$\left(e_{i_1} \cdots e_{i_r}\right)\left(e_{j_1} \cdots e_{j_s}\right) = (-1)^{r-1} c \left(e_{i_2} \cdots e_{i_r}\right)\left(e_{j_2} \cdots e_{j_s}\right)$$

$$\left(e_{j_1} \cdots e_{j_s}\right)\left(e_{i_1} \cdots e_{i_r}\right) = (-1)^{r(s-1)} c \left(e_{i_2} \cdots e_{i_r}\right)\left(e_{j_2} \cdots e_{j_s}\right)$$

Therefore

$$(-1)^{r-1} UV = (-1)^{r(s-1)} VU \Rightarrow (-1)^{r+1} UV = (-1)^{r(s+1)} VU \Rightarrow -UV = (-1)^{rs} VU$$

$$\frac{UV+(-1)^{rs}\,VU}{2}=0.\ \Diamond$$

Corollary 3.4

Let U,V *be multivectors with odd grade. Then*

a. $V \wedge U = -U \wedge V$

b. $U \wedge U = 0$

Proof: These results follow immediately from Equation (3.13). $\Diamond$

Notice that it also follows from Equation (3.13) that the conjugation formula in Equation (3.7) extends to the general wedge product—that is, for arbitrary multivectors U and V

$$\left(sUs^{-1}\right)\wedge\left(sVs^{-1}\right)=s(U \wedge V)s^{-1} \tag{3.14}$$

Exercises

1. Give examples of multivectors U and V where
   a. $V \wedge U \neq -U \wedge V$
   b. $U \wedge U \neq 0$
2. Prove that for arbitrary multivectors U and V: $\left(sUs^{-1}\right)\wedge\left(sVs^{-1}\right)= s(U \wedge V)s^{-1}$.

4 Duality

In n-dimensions, theorems and formulas for subspaces of dimension k often have precise analogues and proofs for subspaces of dimension $n-k$. Moreover, some formulas are symmetric in dimensions k and $n-k$. Therefore we say that subspaces of dimension k are *dual* to subspaces of dimension $n-k$.

To formalize the notion of duality in Clifford algebra, consider a vector space R^n with an orthonormal basis $e_1,\ldots,e_n$. We say that two blades r,s are *dual* if their Clifford product is the pseudoscalar $e_1\cdots e_n$—that is, if $rs=e_1\cdots e_n$. We then extend by linearity the notion of duality to arbitrary multivectors. When s is dual to r, we shall write $s=r^{\perp}$.

To be more concrete and to justify this notation, let σ be a permutation of the integers $1, 2, \ldots, n$. In the standard geometric interpretation of Clifford algebra, a blade $e_{\sigma(1)} \ldots e_{\sigma(k)}$ represents the subspace of R^n of dimension k spanned by the vectors $e_{\sigma(1)}, \ldots, e_{\sigma(k)}$. Under this interpretation, the blade $e_{\sigma(k+1)} \ldots e_{\sigma(n)}$ represents the subspace of R^n of dimension $n - k$ which is the orthogonal complement to the subspace represented by the blade $e_{\sigma(1)} \ldots e_{\sigma(k)}$. Moreover

$$\left(e_{\sigma(1)} \ldots e_{\sigma(k)} \right) \left(e_{\sigma(k+1)} \ldots e_{\sigma(n)} \right) = sign(\sigma) e_1 \cdots e_n.$$

Therefore we write

$$\left(e_{\sigma(1)} \ldots e_{\sigma(k)} \right)^{\perp} = e_{\sigma(k+1)} \ldots e_{\sigma(n)} \quad \text{if } sign(\sigma) = +1$$

$$= -e_{\sigma(k+1)} \ldots e_{\sigma(n)} \quad \text{if } sign(\sigma) = -1$$

to indicate that the subspace represented by $e_{\sigma(k+1)} \ldots e_{\sigma(n)}$ is the orthogonal complement of the subspace represented by $e_{\sigma(1)} \ldots e_{\sigma(k)}$. The sign of the permutation ensures that the two subspaces are oriented consistently.

Notice that this definition of duality is independent of the values the products e_i^2, $i = 1, \ldots, n$. *Duality in Clifford algebra is the same no matter how we define these squares of the basis vectors.* When we shall study Clifford algebras for the dual quaternions, we will investigate two different ways to define these products for the basis vectors. These definitions lead to different algebras, but the notion of duality is formally the same in both algebras.

However, the definition of duality does depend on the dimension n of the initial vector space.

For example

$$(e_1 e_2)^{\perp} = e_3 \qquad n = 3$$

$$(e_1 e_2)^{\perp} = e_3 e_4 \qquad n = 4$$

$$(e_1 e_2)^{\perp} = e_3 e_4 e_5 \qquad n = 5.$$

When we study Clifford algebras for the dual quaternions, we shall see that the 8-dimensional Clifford algebra for the quaternions is embedded in the 16-dimensional Clifford algebras for the dual quaternions. Therefore when we employ duality in these Clifford algebras, we need to be careful to specify with respect to which algebra we are taking the dual.

4.1 Duality in the Quaternion Algebra: Cross Products and Products of Pure Quaternions

In the Clifford algebra for the quaternions—the Clifford algebra generated by $\{1, e_1, e_2, e_3\}$ with e_i^2, $i = 1, 2, 3$ (see Section 2, Example 2.4)—the pseudoscalar is $e_0 = e_1 e_2 e_3$. Hence for blades r, duality is defined in this algebra by the equation $rr^\perp = e_0$. Therefore we have the following dual pairs, where $e_{jk} = e_j e_k$:

$$1^\perp = e_0 \quad \text{and} \quad e_0^\perp = 1$$

$$e_1^\perp = e_{23} \quad \text{and} \quad e_{23}^\perp = e_1$$

$$e_2^\perp = e_{31} \quad \text{and} \quad e_{31}^\perp = e_2$$

$$e_3^\perp = e_{12} \quad \text{and} \quad e_{12}^\perp = e_3$$

In particular, scalars and pseudoscalars are dual. In addition, planes represented by the bivectors e_{jk} are dual to their normal vectors e_i, where $sign(ijk) = 1$. Thus, duality captures orthogonality. Moreover, duality is mediated by multiplication with the pseudoscalar, since

$$e_1^\perp = e_{23} = e_1 e_0 \quad \text{and} \quad e_{23}^\perp = e_1 = -e_{23} e_0$$

$$e_2^\perp = e_{31} = e_2 e_0 \quad \text{and} \quad e_{31}^\perp = e_2 = -e_{31} e_0$$

$$e_3^\perp = e_{12} = e_3 e_0 \quad \text{and} \quad e_{12}^\perp = e_3 = -e_{12} e_0$$

Finally, it is easy to see that the dual of the dual is the identity—that is,

$$\left(r^\perp \right)^\perp = r.$$

One important manifestation of the duality between vectors and bivectors is the duality between the cross product and the wedge product of two vectors. Consider two vectors

$$v_1 = a_1 e_1 + a_2 e_2 + a_3 e_3$$

$$v_2 = b_1 e_1 + b_2 e_2 + b_3 e_3.$$

Then

$$v_1 \wedge v_2 = (a_2 b_3 - a_3 b_2) e_2 e_3 + (a_1 b_3 - a_3 b_1) e_1 e_3 + (a_1 b_2 - a_2 b_1) e_1 e_2.$$

The standard definition of the cross product of two vectors is

$$v_1 \times v_2 = (a_2 b_3 - a_3 b_2) e_1 + (a_3 b_1 - a_1 b_3) e_2 + (a_1 b_2 - a_2 b_1) e_3.$$

Therefore in this model of Clifford algebra, the cross product is dual to the wedge product:

$$v_1 \times v_2 = (v_1 \wedge v_2)^{\perp} \tag{4.1}$$

$$v_1 \times v_2 = -(v_1 \wedge v_2)e_0 = -e_0(v_1 \wedge v_2). \tag{4.2}$$

Thus we can compute the cross product in terms of products already in this Clifford algebra.

The cross product is special to 3-dimensions. Unlike the wedge product, the cross product does not generalize readily to arbitrary dimensions. Also, the cross product is not associative. For these reasons some aficionados of Clifford algebra wish to avoid the cross product altogether and to use instead only the wedge product, especially since the cross product can be computed in terms of the wedge product. However, since we will work in 3-dimensions, we shall embrace the cross product as God's gift to 3-dimensions. For future reference many useful properties of the cross product are included in the Exercises; see too the Appendix to Part I.

There are actually two kinds of cross products in the Clifford algebra for the quaternions: the cross product of two vectors (elements of grade one) and the cross product of two pure quaternions in the quaternion subalgebra (elements of grade 2). There are also two dot products: the dot product of two vectors and the dot product of two pure quaternions. Let $\times$ and $\cdot$ denote the cross product and the dot product of two vectors in the Clifford algebra for the quaternions, and let $\times_q$ and $\cdot_q$ denote the cross product and the dot product of two pure quaternions. These operations are related by duality in the following fashion.

Theorem 4.1

Consider two pure quaternions u, v in the quaternion subalgebra of the Clifford algebra for the quaternions. Then

i. $u \cdot_q v = u^{\perp} \cdot v^{\perp}$

ii. $u \times_q v = -\left(u^{\perp} \times v^{\perp}\right)^{\perp} = v^{\perp} \wedge u^{\perp}$

Proof: Since u, v are two pure quaternions in the quaternion subalgebra:

$$u = u_{32}e_{32} + u_{13}e_{13} + u_{21}e_{21} \leftrightarrow u_{32}i + u_{13}j + u_{21}k$$

$$v = v_{32}e_{32} + v_{13}e_{13} + v_{21}e_{21} \leftrightarrow v_{32}i + v_{13}j + v_{21}k.$$

and by duality

$$u^\perp = -\left(u_{32}e_1 + u_{13}e_2 + u_{21}e_3\right)$$

$$v^\perp = -\left(v_{32}e_1 + v_{13}e_2 + v_{21}e_3\right).$$

i. Therefore

$$u \cdot_q v = u_{32}v_{32} + u_{13}v_{13} + u_{21}v_{21} = u^\perp \cdot v^\perp.$$

ii. Moreover

$$u \times_q v = \left(u_{13}v_{21} - u_{21}v_{13}\right)i + \left(u_{21}v_{32} - u_{32}v_{21}\right)j + \left(u_{32}v_{13} - u_{13}v_{32}\right)k$$

$$\leftrightarrow \left(u_{13}v_{21} - u_{21}v_{13}\right)e_{32} + \left(u_{21}v_{32} - u_{32}v_{21}\right)e_{13} + \left(u_{32}v_{13} - u_{13}v_{32}\right)e_{21}.$$

On the other hand, in the Clifford algebra for the quaternions:

$$u^\perp \times v^\perp = \left(u_{13}v_{21} - u_{21}v_{13}\right)e_1 + \left(u_{21}v_{32} - u_{32}v_{21}\right)e_2 + \left(u_{32}v_{13} - u_{13}v_{32}\right)e_3,$$

so taking the dual we find that

$$-\left(u^\perp \times v^\perp\right)^\perp = \left(u_{13}v_{21} - u_{21}v_{13}\right)e_{32} + \left(u_{21}v_{32} - u_{32}v_{21}\right)e_{13} + \left(u_{32}v_{13} - u_{13}v_{32}\right)e_{21}$$

Comparing the expressions for $u \times_q v$ and $-\left(u^\perp \times v^\perp\right)$, and taking the dual of both sides of Equation (4.1) yields:

$$u \times_q v = -\left(u^\perp \times v^\perp\right)^\perp = v^\perp \wedge u^\perp. \lozenge$$

Consider now the quaternion product of two pure quaternions u, v. It follows from Theorem 4.1 and the definition of the quaternion product that

$$uv = -\left(u \cdot_q v\right) + u \times_q v = -(u^\perp \cdot v^\perp) + v^\perp \wedge u^\perp$$

Thus in the Clifford algebra for the quaternions, the product of two pure quaternions has two terms: a scalar term and a bivector term:

$$-\left(u \cdot_q v\right) = -(u^\perp \cdot v^\perp) \quad \text{(scalar)}$$

$$u \times_q v = v^\perp \wedge u^\perp \quad \text{(bivector = pure quaternion).}$$

Exercise

In Exercises 1–8 use Equation (4.2) along with the formulas for the dot product and wedge products in terms of the Clifford product.

1. Show that
   a. $u \times v = -v \times u$
   b. $u \times u = 0$

2. Show that
   a. $u \cdot (v \times u) = v \cdot (v \times u) = 0$
   b. $(v \times u) \times u = (u \cdot v)u - (u \cdot u)v$

3. Show that
   a. $(u \times v) \times w = (w \cdot u)v - (w \cdot v)u$
   b. $u \times (v \times w) = (u \cdot w)v - (u \cdot v)w$
   c. $u \times (v \times w) + v \times (w \times u) + w \times (u \times v) = 0$

4. Show that
   a. $u \times (v + w) = u \times v + u \times w$
   b. $(v + w) \times u = v \times u + w \times u$

5. Let $u = u_1 e_1 + u_2 e_2 + u_3 e_3$ $v = v_1 e_1 + v_2 e_2 + v_3 e_3$ $w = w_1 e_1 + w_2 e_2 + w_3 e_3$

   and set $\det(u, v, w) = \det \begin{pmatrix} u_1 & v_1 & w_1 \\ u_2 & v_2 & w_2 \\ u_3 & v_3 & w_3 \end{pmatrix}$.

   Show that
   a. $u \wedge v \wedge w = \det(u, v, w)e_0$
   b. $u \cdot (v \times w) = -(u \wedge v \wedge w)e_0$
   c. $u \cdot (v \times w) = \det(u, v, w)$

6. Let u, v be vectors with $u \cdot v = 0$. Show that $|u \times v| = |u||v|$.

7. Let u, v be unit vectors with $u \cdot v = 0$. Show that $v \times (u \times v) = u$.

8. Let u, v be vectors, and let q be a nonzero quaternion. Show that $q(u \times v)q^{-1} = (quq^{-1}) \times (qvq^{-1})$.

9. Let u, v be vectors in the Clifford algebra for the quaternions. Show that
   a. $u^{\perp} \times_q v^{\perp} = v \wedge u$
   b. $e_0 u \times_q e_0 v = v \wedge u$

2

The Plane Model of Clifford Algebra for Dual Quaternions

A plane-based Clifford algebra model for the dual quaternions has been proposed by J. Selig [25] and elaborated on by C. Gunn [11]. We shall call this model *the plane model* for short.

1 Algebra

To develop a Clifford algebra model for the dual quaternions, we start in the plane model with an orthogonal basis g_0, g_1, g_2, g_3 for R^4 so that

$$g_i \cdot g_j = 0 \quad i \neq j,$$

and we set

$$g_0^2 = 0, \quad g_1^2 = g_2^2 = g_3^2 = 1. \tag{1.1}$$

Multiplication in the plane model is associative and distributes through addition, but multiplication is not commutative since

$$g_j g_i = -g_i g_j, \quad i \neq j, i, j = 0, 1, 2, 3.$$

Two important subalgebras reside inside the plane model of Clifford algebra: the quaternions and the dual quaternions. Since $g_1^2 = g_2^2 = g_3^2 = 1$, the entire 8-dimensional Clifford algebra model for quaternions generated by $\{1, g_1, g_2, g_3\}$ is a subalgebra of the plane model of Clifford algebra (see Chapter 1, Section 2, Example 2.4). The quaternions reside inside this model as the even subalgebra generated by $\{1, g_3 g_2, g_1 g_3, g_2 g_1\}$. Furthermore, in this model of Clifford algebra, the dual quaternions are isomorphic to the even subalgebra given by expressions of the form $q_1 + q_2 g_{0123}$, where q_1, q_2 are quaternions—that is $q_1, q_2 \in span(1, g_3 g_2, g_1 g_3, g_2 g_1)$ and $g_{0123} = g_0 g_1 g_2 g_3$. Notice that since $g_0^2 = 0$, it follows that

$$g_{0123}^2 = 0 = \varepsilon^2, \tag{1.2}$$

DOI: 10.1201/9781003398141-15

so we identify g_{0123} with the dual quaternion ε. The square of any blade with the factor g_0 is also zero. We choose g_{0123} to represent ε because

i. g_{0123} lies in the even subalgebra of the plane model generated by the quaternions;

ii. g_{0123} is not used to represent any other geometry;

iii. g_{0123} is compatible with duality (see Section 2.4.2).

To simplify our presentation, we will adopt the following notation:

$$g_{ij} = g_i g_j, \, i \neq j.$$

$$G_0 = g_1 g_2 g_3, \quad G_1 = g_0 g_3 g_2, \quad G_2 = g_0 g_1 g_3, \quad G_3 = g_0 g_2 g_1.$$

$$g_{0123} = g_0 g_1 g_2 g_3,$$

and we shall make use of the following five identities (see Exercise 2):

$$g_i G_i = g_{0123} = -G_i g_i \quad i = 0, 1, 2, 3. \tag{1.3}$$

$$g_i G_0 = g_{jk} = G_0 g_i \quad ijk \text{ is an even permutation of } 123 \tag{1.4}$$

$$g_j G_i = sgn(ijk) g_{0k} = G_i g_j \quad ijk \text{ a permutation of } 123 \tag{1.5}$$

$$G_j G_0 = g_{0j} = -G_0 G_j \quad j \neq 0 \tag{1.6}$$

$$G_0^2 = -1, \quad G_j^2 = 0 \quad j \neq 0 \tag{1.7}$$

With this notation we have the following dictionary for translating between the algebra of dual quaternions and the algebra of the plane model.

TABLE 2.1

Correspondence between Basis Vectors in the Algebra of Dual Quaternions and Blades in the Subalgebra of the Plane Model Isomorphic to the Dual Quaternions

Dual Quaternions	Plane Model
O	1
i	g_{32}
j	g_{13}
k	g_{21}
ε	g_{0123}

The symbol O in the space of dual quaternions plays two roles: algebraic and geometric. Algebraically O represents the identity for multiplication; geometrically O represents the origin. These two roles are split apart in the plane model. In this model of Clifford algebra, 1 is the identity for multiplication, whereas, as we shall see in Section 2.2, G_0 represents the origin. Splitting apart these two roles of O facilitates the separation of operators and operands, one of the advantages of Clifford algebra over the dual quaternions.

Exercises

1. Show that the dual real numbers reside inside the plane model of Clifford algebra as the subalgebra generated by $\{1, g_{0123}\}$.

2. Verify identities (1.3)–(1.7).

3. In this exercise we investigate inverses for the basis multivectors. Show that:

   a. $g_j^{-1} = g_j \quad j \neq 0 \qquad g_{ij}^{-1} = -g_{ij} \quad i, j \neq 0 \qquad G_0^{-1} = -G_0$

   b. $g_0, g_{0j}, G_j, j \neq 0$ are not invertible.

4. Show that:

   a. $G_0 g_0 G_0^{-1} = -g_0$

   b. $G_0 g_j G_0^{-1} = g_j \quad j = 1, 2, 3$

5. Using the results of Exercise 4, show that:

   a. $G_0 g_{0j} G_0^{-1} = -g_{0j} \quad j = 1, 2, 3$

   b. $G_0 g_{ij} G_0^{-1} = g_{ij} \quad i, j = 1, 2, 3$

   c. $G_0 G_0 G_0^{-1} = G_0$

   d. $G_0 G_j G_0^{-1} = -G_j \quad j = 1, 2, 3$

   e. $G_0 g_{0123} G_0^{-1} = -g_{0123}$

2 Geometry

The representation of geometry in the plane model is somewhat exotic. We shall see shortly that in the plane model neither points nor vectors but rather planes are the basic elements; lines are typically represented as the intersection (meet) of two planes, and points are typically represented as the intersection (meet) of three planes.

In the plane model, g_0 does not represent the origin; rather g_0 represents the ideal plane at infinity. The other three basis vectors g_1, g_2, g_3 represent the yz, zx, and xy -coordinate planes. The trivector G_0 (the meet of three planes) represents the origin, while G_1 represents the unit vector in the x-direction; similarly, G_2 represents the unit vector in the y-direction, and G_3 represents the unit vector in the z-direction. In addition, the meet of two planes represents a line. The bivector g_{01} represents the ideal line in the $x = 0$ plane; similarly, g_{02} represents the ideal line in the $y = 0$ plane, and g_{03} represents the ideal line in the $z = 0$ plane. The bivectors g_{32}, g_{13}, g_{21} represent the x-axis, the y-axis, and the z-axis.

These definitions describe some of the geometry associated with this Clifford algebra for the dual quaternions. But why these particular, somewhat peculiar definitions? And how do these definitions relate back to the geometry associated with the dual quaternions described in Part I? Next, we will justify each of these definitions and extend these definitions to representations for arbitrary vectors, points, lines, and planes in 3-dimensions. *The key idea is that multiplication by G_0 maps from the dual quaternion representation of geometry to this Clifford algebra representation of geometry.*

2.1 Planes

To see how one should represent planes in this Clifford algebra model for the dual quaternions, we start with the representation of planes in the space of dual quaternions. The plane with equation $ax + by + cz + d = 0$ is represented by the dual quaternion

$$\tilde{\pi} = ai + bj + ck + dO\varepsilon.$$

The corresponding expression in the plane model of Clifford algebra is (see Table 2.1)

$$\tilde{\pi} = ag_{32} + bg_{13} + cg_{21} + dg_{0123}.$$

Notice, however, that $\tilde{\pi}$ is an element of mixed grade. To generate an element of fixed grade, we can factor

$$\tilde{\pi} = -G_0 \left(ag_1 + bg_2 + cg_3 + dg_0 \right) \tag{2.1}$$

The factor

$$\pi = ag_1 + bg_2 + cg_3 + dg_0 = G_0\tilde{\pi} \tag{2.2}$$

is an element of grade 1, so it is natural to use these grade 1 elements to represent the planes

$$ax + by + cz + d = 0.$$

Thus g_1 represents the yz -plane ($x = 0$), g_2 represents the xz -plane ($y = 0$), and g_3 represents the xy -plane ($z = 0$). The element g_0 represents the plane at infinity. Notice how multiplying by $\pm G_0$ mediates between the dual quaternion and this Clifford algebra representation for planes.

In this representation $v_\pi = ag_1 + bg_2 + cg_3$ corresponds to a normal to the plane π and $d/|v_\pi|$ is the signed distance from the plane π to the origin along this normal direction. When v_π corresponds to a unit normal, the signed distance along this normal from the plane π to the origin is just d. If $d = 0$, then these planes correspond to planes through the origin in the plane-based Clifford algebra model for dual quaternions.

2.2 Points and Vectors

Points with coordinates (p_1, p_2, p_3) are represented by the dual quaternions

$$\tilde{p} = O + (p_1 i + p_2 j + p_3 k)\varepsilon.$$

The corresponding expression in the plane model of Clifford algebra is (see Table 2.1)

$$\tilde{p} = 1 + (p_1 g_{32} + p_2 g_{13} + p_3 g_{21})g_{0123} = 1 + (p_1 g_{01} + p_2 g_{02} + p_3 g_{03}).$$

Notice again that $\tilde{p}$ is an element of mixed grade. To generate an element of fixed grade, we can again factor

$$\tilde{p} = -G_0 (G_0 + p_1 G_1 + p_2 G_2 + p_3 G_3). \tag{2.3}$$

The factor

$$p = G_0 + p_1 G_1 + p_2 G_2 + p_3 G_3 = G_0\tilde{p} \tag{2.4}$$

is an element of grade 3, so it is natural to use these grade 3 elements to represent the points with coordinates (p_1, p_2, p_3). In particular, G_0 represents the point at the origin with coordinates $(0,0,0)$. Notice again how multiplying by $\pm G_0$ mediates between the dual quaternion and this Clifford algebra representation for points.

Since a point is typically represented as a vector emanating from the origin,

$$v_p = p_1 G_1 + p_2 G_2 + p_3 G_3$$

represents the vector from the origin to the point located at (p_1, p_2, p_3). Hence G_1 represents the unit vector along the x-axis, G_2 represents the unit vector along the y-axis, and G_3 represents the unit vector along the z-axis.

Thus, in the plane model points and vectors are both represented by intersections (meets) of three planes. The origin $G_0 = g_1 g_2 g_3 = g_1 \wedge g_2 \wedge g_3$ is the intersection of the three coordinate planes. Vectors are also represented as the intersection (meet) of three planes, but one of these planes is g_0, the plane at infinity.

Exercise

1. Show that for every vector v in the plane model of Clifford algebra, there is a dual quaternion $\tilde{v}$ such that $v = G_0 \tilde{v}$.

2.2.1 Incidence Relation: Point on Plane

Next, we shall show how the incidence relation between a point and a plane in this Clifford algebra model for dual quaternions can be developed directly from the incidence relation in the algebra of dual quaternions. Recall from Part I, Proposition 3.1 that in the space of dual quaternions, a point $\tilde{p}$ lies on the plane $\tilde{\pi}$ if and only if $\tilde{p} \cdot \tilde{\pi} = 0$.

Lemma 2.1

Suppose that $p = G_0 \tilde{p}$ and $\pi = G_0 \tilde{\pi}$. Then

$$\text{a.} \quad \tilde{p} = -G_0 p \text{ and } \tilde{\pi} = -G_0 \pi \tag{2.5}$$

$$\text{b.} \quad \tilde{p}^* = -p G_0 \text{ and } \tilde{\pi}^* = \pi G_0 \tag{2.6}$$

where star denotes the quaternion conjugate.

Proof: Part a follows immediately from Equation (1.7), since $G_0^2 = -1$.

To prove part b, observe that by Equations (1.6) and (1.4)

$$-pG_0 = -(G_0 + p_1G_1 + p_2G_2 + p_3G_3)G_0 = 1 - (p_1g_{32} + p_2g_{13} + p_3g_{21})g_{0123} = \tilde{p}^*$$

$$\pi G_0 = (ag_1 + bg_2 + cg_3 + dg_0)G_0 = -(ag_{32} + bg_{13} + cg_{21}) + dg_{0123} = \tilde{\pi}^* \ \lozenge$$

Proposition 2.2

Let p be a point and π be a plane in the plane model of Clifford algebra, and let $\tilde{p}$ and $\tilde{\pi}$ be the corresponding point and plane in the space of dual quaternions. Then

$$\tilde{p} \cdot \tilde{\pi} = \pi \wedge p. \tag{2.7}$$

Proof: By Lemma 2.1 and the definition of the dot product in the space of dual quaternions (Part I, Equation 2.1)

$$\tilde{p} \cdot \tilde{\pi} = \frac{\tilde{p}\tilde{\pi}^* + \tilde{\pi}\tilde{p}^*}{2} = \frac{(-G_0p)(\pi G_0) + (-G_0\pi)(-pG_0)}{2} = G_0\left(\frac{\pi p - p\pi}{2}\right)G_0.$$

Therefore, by Chapter 1, Proposition 3.3

$$\tilde{p} \cdot \tilde{\pi} = G_0(\pi \wedge p)G_0.$$

But $G_0^2 = -1$, so $G_0^{-1} = -G_0$. Thus by Chapter 1, Equation (3.14)

$$\tilde{p} \cdot \tilde{\pi} = -G_0(\pi \wedge p)G_0^{-1} = (-G_0\pi G_0^{-1}) \wedge (G_0 p G_0^{-1}).$$

Now let

$$\pi = dg_0 + ag_1 + bg_2 + cg_3 = dg_0 + v_\pi$$

$$p = G_0 + p_1G_1 + p_2G_2 + p_3G_3 = G_0 + v_p.$$

Then by Section 1, Exercises 4 and 5

$$-G_0\pi G_0^{-1} = dg_0 - v_\pi$$

$$G_0 p G_0^{-1} = G_0 - v_p.$$

Therefore, since $g_0 \wedge v_p = 0$ and $v_\pi \wedge G_0 = 0$,

$$\tilde{p} \cdot \tilde{\pi} = (dg_0 - v_\pi) \wedge (G_0 - v_p) = dg_0 \wedge G_0 + v_\pi \wedge v_p$$

$$= (dg_0 + v_\pi) \wedge (G_0 + v_p) = \pi \wedge p. \Diamond$$

Theorem 2.3

A point p lies on the plane π if and only if $\pi \wedge p = 0$.

Proof: Let $\tilde{p}$ and $\tilde{\pi}$ be the point and the plane in the space of dual quaternions corresponding to the point p and the plane π in the plane model of Clifford algebra. Then the point p lies on the plane π if and only if the point $\tilde{p}$ lies on the plane $\tilde{\pi}$. Moreover by Part I, Proposition 3.1, the point $\tilde{p}$ lies on the plane $\tilde{\pi}$ if and only if $\tilde{p} \cdot \tilde{\pi} = 0$. But by Equation (2.7) $\tilde{p} \cdot \tilde{\pi} = 0$ if and only if $\pi \wedge p = 0$. Hence the point p lies on the plane π if and only if $\pi \wedge p = 0$. $\Diamond$

Corollary 2.4

Let π_1, π_2, π_3 be three linearly independent planes. Then $p = \pi_1 \wedge \pi_2 \wedge \pi_3$ is the intersection point of the three planes π_1, π_2, π_3.

Proof: This result follows immediately from Theorem 2.3 because

$$\pi_j \wedge p = \pi_j \wedge (\pi_1 \wedge \pi_2 \wedge \pi_3) = 0 \text{ for } j = 1, 2, 3.$$

Therefore, the point $p = \pi_1 \wedge \pi_2 \wedge \pi_3$ lies on all three planes. $\Diamond$

Exercises

1. Let $p = G_0 + xG_1 + yG_2 + zG_3$ be a point and let $\pi = ag_1 + bg_2 + cg_3 + dg_0$ be a plane in the plane model of Clifford algebra.
   a. Using Equations (1.3)–(1.5), show that

   $$\frac{\pi p - p\pi}{2} = (ax + by + cz + d) g_{0123}.$$

   b. Using properties of the wedge product, show that

   $$\pi \wedge p = (ax + by + cz + d) g_{0123}.$$

    c. Conclude from either part *a* or part b that the point p lies on the plane π if and only if

$$\pi \wedge p = 0.$$

2. In this exercise we consider the map $x \to G_0 x G_0^{-1}$.
   a. Show that
      i. $x \to G_0 x G_0^{-1}$ maps planes to planes and points to points
      ii. the point p lies on the plane π if and only if the point $G_0 p G_0^{-1}$ lies on the plane $G_0 \pi G_0^{-1}$
   b. Explain geometrically
      i. how the point $G_0 p G_0^{-1}$ is related to the point p
      ii. the plane $G_0 \pi G_0^{-1}$ is related to the plane π.
   (Hint: See Section 1, Exercise 4.)

2.3 Lines

Lines are typically represented in the space of dual quaternions using either Plucker coordinates or dual Plucker coordinates. Since we are working in the plane model for dual quaternions, it is natural here to use dual Plucker coordinates to represent lines. Moreover, since planes have grade 1 and points have grade 3, we should expect that lines would have grade 2. Finally, since by Corollary 2.4 the wedge product of three linearly independent planes is the intersection point of the three planes, it is again natural to expect that the wedge product of two intersecting planes should represent the intersection line of the two planes.

2.3.1 Lines as the Intersection of Two Planes

Let π_1, π_2 be two intersecting planes. Then $\pi_2 \wedge \pi_1$ generates the dual Plucker coordinates [Part I, Section 3.3.2] of the intersection line of the planes π_1 and π_2.

To see how, let

$$\pi_1 = a_0 g_0 + a_1 g_1 + a_2 g_2 + a_3 g_3 = a_0 g_0 + v_1$$

$$\pi_2 = b_0 g_0 + b_1 g_1 + b_2 g_2 + b_3 g_3 = b_0 g_0 + v_2.$$

Then

$$\pi_2 \wedge \pi_1 = \sum_{i > j} \left(b_i a_j - b_j a_i \right) (g_i \wedge g_j) = \sum_{i > j} \left(b_i a_j - b_j a_i \right) g_{ij}.$$

Thus, the bivector $l = \pi_2 \wedge \pi_1$ has six coefficients relative to the six basis elements g_{ij} of grade 2. These six coefficients correspond to the six dual Plucker coordinates of the line generated by intersecting the two planes π_1 and π_2.

To extract these dual Plucker coordinates for the intersection line from the bivector l, separate the sum for l into two parts: a line at infinity and an affine line:

$$\tilde{u} = (b_1 a_0 - b_0 a_1) g_{10} + (b_2 a_0 - b_0 a_2) g_{20} + (b_3 a_0 - b_0 a_3) g_{30} \quad (\text{line at infinity})$$

$$\tilde{v} = (b_3 a_2 - b_2 a_3) g_{32} + (b_1 a_3 - b_3 a_1) g_{13} + (b_2 a_1 - b_1 a_2) g_{21} \quad (\text{affine line})$$

$$l = \tilde{u} + \tilde{v}.$$

Collect the coefficients of $\tilde{u}$ and $\tilde{v}$ into planes by letting

$$\tilde{u} = u g_0$$

$$u = (a_0 b_1 - b_0 a_1) g_1 + (a_0 b_2 - b_0 a_2) g_2 + (a_0 b_3 - b_0 a_3) g_3$$

$$= a_0 \pi_2 - b_0 \pi_1 = a_0 v_2 - b_0 v_1$$

$$v = \tilde{v} G_0 = (a_2 b_3 - a_3 b_2) g_1 + (a_3 b_1 - a_1 b_3) g_2 + (a_1 b_2 - a_2 b_1) g_3$$

$$= (\pi_1 - a_0 g_0) \times (\pi_2 - b_0 g_0) = v_1 \times v_2.$$

Then by Part I, Section 1.3.3.2, v and u are the dual Plucker coordinates of the intersection line of the two planes π_1, π_2. Moreover, we can express the line l directly in terms of these dual Plucker coordinates, since

$$l = \tilde{u} + \tilde{v} = u g_0 - v G_0 = (a_0 \pi_2 - b_0 \pi_1) g_0 - ((\pi_1 - a_0 g_0) \times (\pi_2 - b_0 g_0)) G_0. \quad (2.8)$$

We can also extract a vector w_l parallel to the line l and a point p_l on the line l from these dual Plucker coordinates. To find a vector w_l parallel to the line l, we can proceed in the following fashion. Notice that the three coefficients of g_{32}, g_{13}, g_{21} in $\tilde{v}$ are the components of the cross product $v_1 \times v_2$. We can extract this cross product as a vector by taking the product of l with g_0. Indeed since $l = \tilde{u} + \tilde{v}$ and $\tilde{u} g_0 = 0$,

$$l g_0 = \tilde{v} g_0 = (a_2 b_3 - a_3 b_2) G_1 + (a_3 b_1 - a_1 b_3) G_2 + (a_1 b_2 - a_2 b_1) G_3 = w_l \quad (2.9)$$

Thus $w_l = lg_0$ is the vector corresponding to $v_1 \times v_2$ perpendicular to the normal vectors of the planes π_1 and π_2 and therefore is a vector parallel to the direction of the intersection line l.

To find a point p_l on the line l, we can intersect three planes. Consider again the plane

$$v = \tilde{v}G_0 = (a_2 b_3 - a_3 b_2)g_1 + (a_3 b_1 - a_1 b_3)g_2 + (a_1 b_2 - a_2 b_1)g_3$$

The normal vector to v points in the direction $v_1 \times v_2$ parallel to the line l. Hence the normal vectors to the three planes π_1, π_2, v are linearly independent, so by Corollary 2.4 the point

$$p_l = \pi_2 \wedge \pi_1 \wedge v$$

lies on all three planes and therefore lies on the line $l = \pi_2 \wedge \pi_1$. To simplify this triple product, recall that

$$\pi_2 \wedge \pi_1 = l = \tilde{u} + \tilde{v}$$

$$\tilde{v} = (b_3 a_2 - b_2 a_3)g_3 \wedge g_2 + (b_1 a_3 - b_3 a_1)g_1 \wedge g_3 + (b_2 a_1 - b_1 a_2)g_2 \wedge g_1$$

$$v = \tilde{v}G_0 = (a_2 b_3 - a_3 b_2)g_1 + (a_3 b_1 - a_1 b_3)g_2 + (a_1 b_2 - a_2 b_1)g_3$$

Now expanding and simplifying yields (see Exercise 4)

$$\tilde{v} \wedge v = -|v|^2 G_0$$

Therefore

$$p_l = (\pi_2 \wedge \pi_1) \wedge v = l \wedge v = \tilde{v} \wedge v + \tilde{u} \wedge v$$

$$= -|v|^2 G_0 + \tilde{u} \wedge v \equiv G_0 - \frac{\tilde{u} \wedge v}{|v|^2}. \tag{2.10}$$

Notice three things. First, if the planes π_1 and π_2 both pass through the origin, then $a_0 = b_0 = 0$, so $\tilde{u} = 0$. Therefore, in this case, as expected, the point p_l on the line l lies at the origin G_0. Second, $\tilde{u} \wedge v$ is a vector (grade 3), so the term $\dfrac{\tilde{u} \wedge v}{|v|^2}$ measures the displacement of the point p_l from the origin G_0. Third, in the definition of the plane v we have chosen the coefficient of g_0 to be zero. In fact, we can choose this coefficient to be any arbitrary scalar value and we would still generate a point on the line l; different scalar values would simply generate different points on the same line (see Exercise 5). Thus, there is one free parameter corresponding to the one parameter family of points along the line l.

Once we have one point p_l on the line l and a vector w_l parallel to the line l, we can find arbitrarily many points p on the line l, using the formula $p = p_l + \lambda w_l$, where λ is a real number.

Exercises

1. Consider the two lines

$$\tilde{v} = (b_3 a_2 - b_2 a_3) g_{32} + (b_1 a_3 - b_3 a_1) g_{13} + (b_2 a_1 - b_1 a_2) g_{21}$$

$$\tilde{u} = (b_1 a_0 - b_0 a_1) g_{10} + (b_2 a_0 - b_0 a_2) g_{20} + (b_3 a_0 - b_0 a_3) g_{30}$$

Show that:
   a. The line $\tilde{v}$ is the intersection of two planes passing through the origin.
   b. The line $\tilde{u}$ is a line at infinity.

2. Consider the planes u, v defined in the text by setting:

$$u = (a_0 b_1 - b_0 a_1) g_1 + (a_0 b_2 - b_0 a_2) g_2 + (a_0 b_3 - b_0 a_3) g_3 = a_0 \pi_2 - b_0 \pi_1$$

$$v = (a_2 b_3 - a_3 b_2) g_1 + (a_3 b_1 - a_1 b_3) g_2 + (a_1 b_2 - a_2 b_1) g_3$$

$$= (\pi_1 - a_0 g_0) \times (\pi_2 - b_0 g_0)$$

Show that $u \cdot v = 0$ in two ways:
   a. by direct computation
   b. by using some standard properties of determinants.

3. Show that if a point lies on the two planes corresponding to π_1 and π_2, then the point also lies on the plane $u = a_0 \pi_2 - b_0 \pi_1$.

4. Consider the line $\tilde{v}$ and the plane v where

$$\tilde{v} = (b_3 a_2 - b_2 a_3)(g_3 \wedge g_2) + (b_1 a_3 - b_3 a_1)(g_1 \wedge g_3) + (b_2 a_1 - b_1 a_2)(g_2 \wedge g_1)$$

$$v = \tilde{v} G_0 = (a_2 b_3 - a_3 b_2) g_1 + (a_3 b_1 - a_1 b_3) g_2 + (a_1 b_2 - a_2 b_1) g_3$$

a. Show that $\tilde{v} \wedge v = -|v|^2 G_0$ in two ways: by invoking Chapter 1, Equation (3.11), (3.12) or by invoking Chapter 1, Equation (3.13).
b. Explain geometrically why the line $\tilde{v}$ and the plane v must intersect at the origin.

5. Consider the computation of the point p_l on the line l. Show that if we choose the plane v with a g_0 component, then the new point p_0 lies on the same line l as the point p_l. (Hint: Show that the vector $p_0 - p_l$ is parallel to the direction vector of the line l.)

6. Consider two planes

$$\pi_1 = a_0 g_0 + a_1 g_1 + a_2 g_2 + a_3 g_3$$

$$\pi_2 = b_0 g_0 + b_1 g_1 + b_2 g_2 + b_3 g_3.$$

Suppose that π_1 and π_2 are parallel planes. Show that
a. $\pi_2 \wedge \pi_1 = (b_1 a_0 - b_0 a_1) g_{10} + (b_2 a_0 - b_0 a_2) g_{20} + (b_3 a_0 - b_0 a_3) g_{30}$
b. $\pi_2 \wedge \pi_1$ is a line at infinity.

7. Show that if l represents a line as the intersection of two planes, then cl represents the same line for all constants $c \neq 0$.

8. Show that, up to constant multiples, the bivector representing a line as the meet of two planes is independent of the choice of the two planes whose intersection is the line. That is, show that if π_1, π_2 and ρ_1, ρ_2 intersect in the same line, then there is a constant $\lambda \neq 0$ such that $\rho_2 \wedge \rho_1 = \lambda(\pi_2 \wedge \pi_1)$. (Hint: Express ρ_1, ρ_2 in terms of π_1, π_2.)

2.3.2 Lines as Bivectors

We have seen that bivectors can represent lines as the intersection of two planes. But not all bivectors represent lines. Recall that lines are represented by dual quaternions of the form

$$l = \tilde{v} + \tilde{u}\varepsilon,$$

where $\tilde{v}$ and $\tilde{u}$ are pure quaternions and $\tilde{u} \cdot \tilde{v} = 0$. The corresponding expressions in the plane model of Clifford algebra are

$$\tilde{v} = l_{32} g_{32} + l_{13} g_{13} + l_{21} g_{21}$$

$$\tilde{u}\varepsilon = (l_{01} g_{32} + l_{02} g_{13} + l_{03} g_{21}) g_{0123} = l_{01} g_{01} + l_{02} g_{02} + l_{03} g_{03},$$

so

$$l = \tilde{v} + \tilde{u}\varepsilon = l_{32}g_{32} + l_{13}g_{13} + l_{21}g_{21} + l_{01}g_{01} + l_{02}g_{02} + l_{03}g_{03}$$

$$l_{01}l_{32} + l_{02}l_{13} + l_{03}l_{21} = \tilde{u} \cdot \tilde{v} = 0.$$

Therefore, if a bivector l represents a line (or equivalently the intersection of two planes), then we must have

$$l_{01}l_{32} + l_{02}l_{13} + l_{03}\ l_{21} = 0.$$

Hence if $l_{01}l_{32} + l_{02}l_{13} + l_{03}\ l_{21} \neq 0$, then the bivector l does not represent a line.

But if $l_{01}l_{32} + l_{02}l_{13} + l_{03}\ l_{21} = 0$, then we can always find two planes π_1, π_2 such that $l = \pi_2 \wedge \pi_1$. To find two such planes (not necessarily unique), by Equation (2.8) we must find two planes

$$\pi_1 = a_0 g_0 + a_1 g_1 + a_2 g_2 + a_3 g_3 = a_0 g_0 + v_1$$

$$\pi_2 = b_0 g_0 + b_1 g_1 + b_2 g_2 + b_3 g_3 = b_0 g_0 + v_2$$

such that

$$\tilde{u}\varepsilon = u g_0$$

$$u = -\left(l_{01}g_1 + l_{02}g_2 + l_{03}g_3\right) = a_0 \pi_2 - b_0 \pi_1 = a_0 v_2 - b_0 v_1$$

$$v = \tilde{v} G_0 = l_{32}g_1 + l_{13}g_2 + l_{21}g_3 = (\pi_1 - a_0 g_0) \times (\pi_2 - b_0 g_0) = v_1 \times v_2$$

$$l = \tilde{v} + \tilde{u}\varepsilon = u g_0 - v G_0$$

If $u = 0$, then since π_1 and π_2 are necessarily linearly independent, it follows that $a_0 = b_0 = 0$. Hence we need only find (normal) vectors whose cross product is v. Let w be any unit normal vector perpendicular to v. Then $v, w, v \times w$ are mutually orthogonal vectors, so $w \times (v \times w) = v$. Thus when $u = 0$, we can simply choose $\pi_1 = w$ and $\pi_2 = v \times w$.

It remains to treat the case where $u \neq 0$. By normalizing the planes π_1, π_2, we can choose $a_0, b_0 = 0, 1$ with the condition that a_0, b_0 are not both zero. Notice that v and u have no g_0 term, so v and u are (normal) vectors. Therefore, we need only find (normal) vectors w_1, w_2 such that

$$a_0 w_2 - b_0 w_1 = u$$

$$w_1 \times w_2 = v.$$

Then we can set

$$\pi_1 = a_0 g_0 + w_1$$

$$\pi_2 = b_0 g_0 + w_2.$$

Since the wedge product and the cross product are related by multiplication with G_0 (Chapter 1, Equation 4.2), we would then have

$$w_2 \wedge w_1 = (w_2 \times w_1)G_0 = -(w_1 \times w_2)G_0 = -vG_0$$

$$\pi_2 \wedge \pi_1 = (b_0 g_0 + w_2) \wedge \left(a_0 g_0 + w_1\right)$$

$$= \left(a_0 w_2 - b_0 w_1\right) \wedge g_0 + w_2 \wedge w_1 = ug_0 - vG_0 = l.$$

Now notice that

$$u \cdot v = -\left(l_{01} l_{32} + l_{02} l_{13} + l_{03}\ l_{21}\right) = 0.$$

But whenever $u \cdot v = 0$ we can always construct two such vectors w_1, w_2 using the following theorem.

Theorem 2.5

Suppose that $u, v \neq 0$ and that $v \cdot u = 0$. Set

- $w_1 = \alpha \dfrac{u}{|u|} - a_0 \dfrac{|v|}{|u|} \left(\dfrac{v \times u}{|v||u|} \right)$

- $w_2 = \gamma \dfrac{u}{|u|} - b_0 \dfrac{|v|}{|u|} \left(\dfrac{v \times u}{|v||u|} \right)$

where α, γ are constants that satisfy the constraint $a_0\gamma - b_0\alpha = |u|$. Then

   i. $a_0 w_2 - b_0 w_1 = u$

   ii. $w_1 \times w_2 = v.$

Proof: See Lemma 3.3 from Part I, Section 1.3.3.2. ◊

In our final result for this section we recall how to construct a point p_l on the line l and a vector w_l parallel to the line l.

Theorem 2.6

Let $l = l_{32}g_{32} + l_{13}g_{13} + l_{21}g_{21} + l_{01}g_{01} + l_{02}g_{02} + l_{03}g_{03}$ be a bivector representing a line. Set

- $\tilde{u} = l_{01}g_{01} + l_{02}g_{02} + l_{03}g_{03}$

- $\tilde{v} = l_{32}g_{32} + l_{13}g_{13} + l_{21}g_{21}$

- $v = \tilde{v}G_0$

Then

   i. $p_l = G_0 - \dfrac{\tilde{u} \times v}{|v|^2}$ *is a point on the line l.*

   ii. $w_l = lg_0 = \tilde{v}g_0$ *is a vector parallel to the line l.*

Thus, the direction of a line l depends only on quaternion term $\tilde{v} = l_{32}g_{32} + l_{13}g_{13} + l_{21}g_{21}$. Two lines that share the same quaternion term $\tilde{v}$ are necessarily parallel (see too Exercise 5).

Proof: This result follows from Equations (2.9) and (2.10) in Section 2.3.1 and the fact that every bivector representing a line can be written as the wedge product of two planes. ◊

Remark 2.7

In Theorem 2.6 we have tacitly assumed that $\tilde{v} = l_{32}g_{32} + l_{13}g_{13} + l_{21}g_{21} \neq 0$. If $\tilde{v} = 0$, then the bivector $l = \tilde{u} = l_{01}g_{01} + l_{02}g_{02} + l_{03}g_{03} = g_0\left(l_{01}g_1 + l_{02}g_2 + l_{03}g_3\right)$ is the intersection of an affine plane and a plane at infinity, which is a line at infinity.

Exercises

1. Let u, v be normal vectors. Show that the bivector $l = v \wedge u$ represents a line through the origin parallel to the normal vector $u \times v$.

2. Consider a bivector

$$l = l_{32}g_{32} + l_{13}g_{13} + l_{21}g_{21} + l_{01}g_{01} + l_{02}g_{02} + l_{03}g_{03}.$$

   Let

$$\bar{v} = l_{01}g_{01} + l_{02}g_{02} + l_{03}g_{03}$$

$$\tilde{u} = l_{32}g_{32} + l_{13}g_{13} + l_{21}g_{21}.$$

   Show that l represents a line if and only if $\tilde{v} \wedge \tilde{u} = 0$.

3. Show that:
   a. g_{32} represents the x-axis
   b. g_{13} represents the the y-axis
   c. g_{21} represents the z-axis.

4. Show that:
   a. g_{01} represents the ideal line in the $x = 0$ plane
   b. g_{02} represents the ideal line in the $y = 0$ plane
   c. g_{03} represents the ideal line in the $z = 0$ plane.

5. Consider two affine lines l, l' represented by the bivectors

$$l = l_{32}g_{32} + l_{13}g_{13} + l_{21}g_{21} + l_{01}g_{01} + l_{02}g_{02} + l_{03}g_{03}$$

$$l' = l'_{32}g_{32} + l'_{13}g_{13} + l'_{21}g_{21} + l'_{01}g_{01} + l'_{02}g_{02} + l'_{03}g_{03}.$$

   Show that the lines l, l' are parallel if and only if there is a constant $\lambda \neq 0$ such that

$$l'_{32}g_{32} + l'_{13}g_{13} + l'_{21}g_{21} = \lambda \left(l_{32}g_{32} + l_{13}g_{13} + l_{21}g_{21} \right).$$

6. Let l be a bivector representing a line.
   a. Show that $G_0 l G_0^{-1}$ is also a bivector representing a line.
   b. Explain geometrically how the line $G_0 l G_0^{-1}$ is related to the line l.

   (Hint: See Section 1, Exercise 5.)

7. In this exercise we consider the map $x \to xG_0$. Show that:

    a. $x \to xG_0$ maps between planes through the origin and lines through the origin perpendicular to these planes – that is, lines through the origin parallel to the normal vectors.

    b. $x \to xG_0$ maps between vectors and lines at infinity.

2.3.3 Incidence Relations for Lines

Given an affine line represented by a bivector l, five problems naturally arise:

1. Find two planes (not necessarily unique) whose intersection is the line l.
2. Determine whether a point p lies on the line l.
3. Determine whether the line l lies on a plane π.
4. Find the intersection point p of the line l and a plane π.
5. Determine whether the line l intersects another line l' in 3-dimensions.

We have already seen in the preceding section how to solve the first problem. We shall see shortly that we can use the solution to the first problem to solve the other four problems.

To determine if a point p lies on the line l, first find two planes π_1 and π_2 such that $l = \pi_2 \wedge \pi_1$. Then p lies on l if and only if p lies on both π_1 and π_2. Therefore, by Theorem 2.3, p lies on l if and only if $p \wedge \pi_1 = 0$ and $p \wedge \pi_2 = 0$. For a more direct solution for determining if a point p lies on a line represented by a bivector l without finding two such planes, see Exercise 4. For a more compact formula, see Corollary 2.12 in Section 2.4.3.

Next, we show how to determine if a line l lies on a plane π. Moreover, if the line l does not lie on the plane π, we show how to compute the intersection point p of the line l and the plane π.

Theorem 2.8

Let l be a line and let π be a plane.

1. *If the line l lies on the plane π, then $\pi \wedge l = 0$.*
2. *If the line l does not lie on the plane π, then $p = \pi \wedge l$ is the point of intersection of the line l and the plane π.*

Proof:

1. Suppose that the line l lies on the plane π. Then there is another plane τ such that l is the intersection of π and τ. Therefore $l = \pi \wedge \tau$, so $\pi \wedge l = \pi \wedge \pi \wedge \tau = 0$.
2. If the line l does not lie on the plane π, then we can find two planes $\pi_1, \pi_2 \neq \pi$ such that $l = \pi_2 \wedge \pi_1$. Therefore by Corollary 2.4, the point $p = \pi \wedge l = \pi \wedge \pi_2 \wedge \pi_1$ lies on all three planes, so $p = \pi \wedge l$ is the point of intersection of the line l and the plane π. Note that the point p might lie at infinity—that is, is a vector, if the line l is parallel to the plane π. $\Diamond$

We close this section by showing how to determine whether or not two lines l_1, l_2 in 3-dimensions intersect.

Theorem 2.9

Let l_1 and l_2 be bivectors representing two distinct lines. Then the lines l_1 and l_2 intersect if and only if $l_1 \wedge l_2 = 0$.

Proof: Let $l_1 = \pi_1 \wedge \pi_2$ and $l_2 = \pi_3 \wedge \pi_4$, If $l_1 \wedge l_2 = 0$, then $\pi_1 \wedge \pi_2 \wedge \pi_3 \wedge \pi_4 = 0$, so the point $p = \pi_1 \wedge \pi_2 \wedge \pi_3$ lies on all four planes; hence the point p lies on both l_1 and l_2. Conversely, if a point p lies on both l_1 and l_2, then $p \wedge \pi_j = 0$ for $j = 1, 2, 3, 4$. Hence the point p certainly lies on the planes π_1, π_2, π_3, so $p = \pi_1 \wedge \pi_2 \wedge \pi_3$. But the point p also lies on the plane π_4, so $l_1 \wedge l_2 = \pi_1 \wedge \pi_2 \wedge \pi_3 \wedge \pi_4 = p \wedge \pi_4 = 0$. $\Diamond$

Exercises

1. Recall that every vector v can be expressed as $v = v_1 G_1 + v_2 G_2 + v_3 G_3$.
   a. Show that every vector is the intersection of an affine line and the plane at infinity.
   b. Conclude that every vector is the intersection of three planes.

2. Let p be a point and let l be a bivector representing a line. Explain why $p \wedge l = 0$ does not prove that the point p lies on the line l.

3. Let p be a point and let l be a line. Show that the point p lies on the line l if and only if the point $G_0 p G_0^{-1}$ lies on the line $G_0 l G_0^{-1}$. (Hint: See Section 2.2.1, Exercise 2.)

4. Consider a line l and a point p where:

$$l = l_{32} g_{32} + l_{13} g_{13} + l_{21} g_{21} + l_{01} g_{01} + l_{02} g_{02} + l_{03} g_{03}$$

$$p = G_0 + p_1 G_1 + p_2 G_2 + p_3 G_3 = G_0 + g_0 v_p.$$

Let

$$v = l_{32} g_{32} + l_{13} g_{13} + l_{21} g_{21}$$

$$u = l_{01} g_{01} + l_{02} g_{02} + l_{03} g_{03}.$$

Show that the point p lies on the line l if and only if $\left((v G_0) \wedge (v_p G_0)\right) g_{0123} = u.$

(Hint: Start with the dual quaternion formula for determining whether a point lies on a line in Part I, Section 1.3.3.1, Exercise 4 and convert this formula from dual quaternions to Clifford algebra.)

5. Consider two distinct lines $l = \pi_2 \wedge \pi_1$ and $l' = \pi_4 \wedge \pi_3$, Show that if the lines l, l' intersect, then the intersection point is $p = \pi_3 \wedge \pi_2 \wedge \pi_1$, where the three planes π_1, π_2, π_3 are chosen to be linearly independent.

2.4 Duality

Duality in the Clifford algebra associated to a vector space of dimension n captures relationships between subspaces of dimension k and subspaces of dimension $n - k$ (see Chapter 1, Section 4). The plane model of Clifford algebra focuses on geometry in 3-dimensions. In 3-dimensions, subspaces of dimension k are dual to subspaces of dimension $3 - k$. Theorems and formulas for subspaces of dimension k often have precise analogues and parallel proofs for subspaces of dimension $3 - k$. Moreover, some formulas are symmetric in dimensions k and $3 - k$. For example, a point p lies on a plane π if and only if $\pi \wedge p = 0$. This formula is symmetric in the point p and the plane π—that is, symmetric in dimensions 0 and $3 - 0 = 3$. Here we briefly review duality in the quaternion subalgebra and then go on to investigate duality in the plane model of Clifford algebra.

2.4.1 Duality in the Quaternion Subalgebra

We have already encountered duality in the quaternion subalgebra of the plane model—the subalgebra generated by $\{1, g_1, g_2, g_3\}$. For quaternions two blades r, s are dual if $rs = G_0$. Since for every blade s in this subalgebra $s^2 = \pm 1$, it follows that two blades r and s are dual if and only if $r = \pm G_0 s$. Moreover $G_0^2 = -1$, so if $r = G_0 s$, then $s = -G_0 r$. Thus in the quaternion subalgebra duality is mediated by multiplication with $\pm G_0$. We invoked this notion of duality in Chapter 1, Section 4.1 to show that the cross product of two vectors is dual to their wedge product, and we then used this duality to investigate the algebraic properties of the cross product.

Multiplication with G_0 also maps the dual quaternion representations for planes, points, and vectors to the Clifford algebra representations for planes, points, and vectors. So in this sense the Clifford algebra representations for planes, points and vectors are dual to the dual quaternion representations for planes, points, and vectors. This duality simplifies the analysis of geometry in the plane model of Clifford algebra by allowing us to apply formulas previously derived in the algebra of dual quaternions. Notice, however, that multiplication with G_0 does not map the dual quaternion representation for lines (dual Plücker coordinates) to the Clifford algebra representation for lines: multiplication with G_0 meditates only between the dual quaternion $\tilde{v}$ and Clifford algebra representations $v = \tilde{v} G_0 = G_0 \tilde{v}$ of the direction vector of the line (see Section 2.3.2). To capture the other component of the line, we need another form of duality.

2.4.2 Duality in the Plane Model

In the plane model of Clifford algebra the pseudoscalar is g_{0123}. Thus in this algebra two blades r, s are said to be *dual* if $rs = g_{0123}$. In the quaternion subalgebra, duality is mediated in both directions by multiplication with the pseudoscalar G_0. But in the plane model of Clifford algebra, pseudoscalars are not invertible since $g_{0123}^2 = 0$. Therefore we need to be more careful when we compute duality using pseudoscalars in the plane model of Clifford algebra.

We shall define duality in the plane model for the basis multivectors and then extend by linearity. As in Chapter 1, Section 4 we adopt the notation $s = r^\perp$ to indicate that s is dual to r. For the basis multivectors when s is dual to r, we want $rs = g_{0123}$, so we define duality for the basis multivectors in dual pairs:

$$1^\perp = g_{0123} \quad and \quad g_{0123}^\perp = 1 \tag{2.11}$$

$$g_j^\perp = G_j \quad and \quad G_j^\perp = -g_j \tag{2.12}$$

$$g_{ij}^\perp = -g_{0k} \quad and \quad g_{0k}^\perp = -g_{ij} \quad (ijk \text{ is an even permutation of } 321) \tag{2.13}$$

For each of these dual pairs r, s either $r = \pm s g_{0123}$ or $s = \pm r g_{0123}$, so we have tried to maintain the spirit of duality as multiplication by a pseudoscalar. But both equalities cannot hold simultaneously because $g_{0123}^2 = 0$. Notice that in some cases to find the dual we multiply by a pseudoscalar, while in other cases we remove a pseudoscalar. Due to the minus signs,

$$rr^\perp = g_{0123} \quad \text{for all grades} \tag{2.14}$$

$$(r^\perp)^\perp = r \quad \text{for even grades } (\text{dual quaternions}) \tag{2.15}$$

$$(r^\perp)^\perp = -r \quad \text{for odd grades} \tag{2.16}$$

These formulas for duality in the plane model mimic, but are not identical to, duality formulas for dual quaternions: some signs are changed. For example, for dual quaternions,

$$i^\# = i\varepsilon, \quad j^\# = j\varepsilon, \quad k^\# = k\varepsilon;$$

the equivalent formulas in the plane model would be

$$g_{ij}^\perp = g_{ij} g_{0123} = g_{0k} \neq -g_{0k} \quad (ijk \text{ is an even permutation of 321}).$$

The duality formulas for the basis elements in the space of dual quaternions are chosen so that

$$q \cdot q^\# = \varepsilon.$$

In contrast, the duality formulas for the basis multivectors in the plane model are chosen so that

$$rr^\perp = g_{0123}$$

These different constraints account for the differences in the formulas for duality in the space of dual quaternions and in the plane model of Clifford algebra.

Nevertheless, as in the space of dual quaternions, planes and their normal vectors are dual in the plane model of Clifford algebra. In fact, the most important example of duality in the plane model is that, up to sign, g_j and G_j are dual for $j = 1, 2, 3$. Thus, in general, vectors are dual to planes through the origin. Since the vector G_j is orthogonal to the plane g_j with the correct orientation, this duality once again justifies our choice of the superscript $\perp$ to represent duality. Also, up to sign, the origin G_0 and the plane at infinity g_0 are dual, and affine lines and lines at infinity are dual.

Notice too that the second component of the dual Plucker coordinates of a line represented in dual quaternions is now, up to sign, dual to the representation of this component in the plane model of Clifford algebra:

$$\tilde{u}\varepsilon = \left(l_{01}g_{32} + l_{02}g_{13} + l_{03}g_{21}\right)g_{0123} = l_{01}g_{01} + l_{02}g_{02} + l_{03}g_{03}.$$

We shall apply this duality in the plane model of Clifford algebra to great effect to construct bivectors representing lines as the joins of two points (Section 2.4.3), and to compute perspective and pseudo-perspective projections using the translation rotor (Section 3.4).

In addition, the duality between vectors and planes facilitates the definition of lengths and angles for vectors in the plane model. Consider two vectors

$$u = u_1 G_1 + u_2 G_2 + u_3 G_3$$

$$v = v_1 G_1 + v_2 G_2 + v_3 G_3$$

and let θ be the angle between u and v. Then

$$|u|^2 = u^\perp \cdot u^\perp = (u_1 g_1 + u_2 g_2 + u_3 g_3)(u_1 g_1 + u_2 g_2 + u_3 g_3) = u_1^2 + u_2^2 + u_3^2$$

$$\cos(\theta) = \frac{u^\perp \cdot v^\perp}{|u| |v|}.$$

Exercises

1. Show that even for basis multivectors, in general
   a. $(r \wedge s)^\perp \neq r^\perp \wedge s^\perp$
   b. $(r \wedge s)^\perp \neq s^\perp \wedge r^\perp$
   c. $r \wedge s^\perp \neq r^\perp \wedge s$.

2. Let u and v be normal vectors. Show that:
   a. $(u \cdot v)^\perp = v \wedge u^\perp$.
   b. $(u \wedge v)^\perp = g_0(u \times v)$.

3. Let π be the plane with normal vector v passing through the point p. Show that:
   a. $(p \wedge v)^\perp$ is a scalar
   b. $\pi = v - (p \wedge v)^\perp g_0$.

4. Let p be a point and let π be a plane. Show that:

    a. $\pi^{\perp} \wedge p^{\perp} = -(\pi \wedge p)$. (Hint: See Section 2.2.1, Exercise 1b.)

    b. $\left(p_1^{\perp} \wedge p_2^{\perp} \wedge p_3^{\perp}\right)^{\perp}$ is the plane containing the three noncollinear points p_1, p_2, p_3. (Hint: Use part a.)

5. Let π be a plane with a unit normal vector v and containing a point q, and let p be an arbitrary point. Show that:

    a. $dist(p, \pi) = \left|(p - q)^{\perp} \cdot v\right|$

    b. $dist(p, \pi) = \left|(\pi \wedge p)^{\perp}\right|$.

6. Let l be a bivector representing a line.

    a.   Show that $l^{\perp}$ is also a bivector representing a line.

    b.   How is the line $l^{\perp}$ related to the line l?

    (Compare to Part I, Section 1.3.4, Exercise 8.)

7. Consider the bivector
$$l = l_{32} g_{32} + l_{13} g_{13} + l_{21} g_{21} + l_{01} g_{01} + l_{02} g_{02} + l_{03} g_{03}, \text{ and let}$$

$$v = l_{32} g_1 + l_{13} g_2 + l_{21} g_3$$

$$u = l_{01} g_1 + l_{02} g_2 + l_{03} g_3$$

Show that:

    a. $v = -(g_0 l)^{\perp}$

    b. $u = (g_0 l^{\perp})^{\perp}$

8. Let l be a bivector representing a line with direction vector $v = l g_0$ and containing a point q, and let p be arbitrary point. Show that:

    a. $dist^2(p, l) = (p - q)^{\perp} \cdot (p - q)^{\perp} - \dfrac{\left((p - q)^{\perp} \cdot v^{\perp}\right)^2}{v^{\perp} \cdot v^{\perp}}$

    b. $dist^2(p, l) = \dfrac{\left|(p - q)^{\perp} \times v^{\perp}\right|}{\left|v^{\perp}\right|}$.

2.4.3 Lines as the Join of Two Points

Two points determine a line. But in the plane model for dual quaternions, the bivector representing a line is constructed from the intersection (meet) of two planes (see Section 2.3.1). Suppose, however, that we are given two points; what then is the bivector representing the line joining these two points?

Since by Equation (2.12) the dual of a point is a plane, one might expect that the line joining two points p, q is given by the intersection of their two dual planes—that is, by $q^{\perp} \wedge p^{\perp}$, where the order of the planes $p^{\perp}$ and $q^{\perp}$ is reversed as in Section 2.3.1 to ensure that the join is oriented from p to q. This formula is almost correct, but not quite. In fact, the line joining p to q is the dual of the line $q^{\perp} \wedge p^{\perp}$. Therefore, we define the join of two points by setting

$$p \vee q = \left(q^{\perp} \wedge p^{\perp}\right)^{\perp}. \tag{2.17}$$

The join $p \vee q$ is an antisymmetric, bilinear function (see Exercise 1). Hence, up to constant multiples, the bivector representing the join of two points is independent of the choice of the two points on the line (see Exercise 2).

Theorem 2.10

The bivector $l = \left(q^{\perp} \wedge p^{\perp}\right)^{\perp}$ represents the line joining the points p and q.

Proof: Consider two points

$$p = G_0 + p_1 G_1 + p_2 G_2 + p_3 G_3 = G_0 + v_p$$

$$q = G_0 + q_1 G_1 + q_2 G_2 + q_3 G_3 = G_0 + v_q$$

By Equation (2.12)

$$p^{\perp} = -(g_0 + p_1 g_1 + p_2 g_2 + p_3 g_3) = -g_0 + v_p^{\perp}$$

$$q^{\perp} = -(g_0 + q_1 g_1 + q_2 g_2 + q_3 g_3) = -g_0 + v_q^{\perp}$$

$$q^{\perp} \wedge p^{\perp} = (p_1 - q_1) g_{01} + (p_2 - q_2) g_{02} + (p_3 - q_3) g_{03}$$

$$+ (q_3 p_2 - q_2 p_3) g_{32} + (q_1 p_3 - q_3 p_1) g_{13} + (q_2 p_1 - q_1 p_2) g_{21}$$

Therefore, by Equation (2.13),

$$l = \left(q^{\perp} \wedge p^{\perp}\right)^{\perp} = (q_1 - p_1)g_{32} + (q_2 - p_2)g_{13} + (q_3 - p_3)g_{21}$$
$$+ (q_2 p_3 - q_3 p_2)g_{01} + (q_3 p_1 - q_1 p_3)g_{02} + (q_1 p_2 - q_2 p_1)g_{03}.$$

Let

$$v = l_{01}g_1 + l_{02}g_2 + l_{03}g_3 = (q_2 p_3 - q_3 p_2)g_1 + (q_3 p_1 - q_1 p_3)g_2 + (q_1 p_2 - q_2 p_1)g_3$$

$$u = l_{32}g_1 + l_{13}g_2 + l_{32}g_3 = (q_1 - p_1)g_1 + (q_2 - p_2)g_2 + (q_3 - p_3)g_3.$$

Then

$$v = v_q^{\perp} \times v_p^{\perp}$$

$$u = v_p^{\perp} - v_q^{\perp}$$

$$l = g_0 v - G_0 u. \tag{*}$$

Now we need to show three things:

   i. The bivector $l = \left(q^{\perp} \wedge p^{\perp}\right)^{\perp}$ represents a line.
  ii. The line represented by the bivector l is parallel to the line joining p to q.
 iii. There is a point on the line l that is also on the line joining p to q.

We shall prove each statement in turn.

   i. It is straightforward to verify (see Exercise 3) that

$$l_{01}l_{32} + l_{02}l_{13} + l_{03}\,l_{21} = v \cdot u = 0.$$

Therefore, by Section 2.3.2, the bivector $l = \left(q^{\perp} \wedge p^{\perp}\right)^{\perp}$ represents a line.

ii. By Section 2.3.2, the direction of the line l is given by

$$lg_0 = (g_0 v - G_0 u) g_0 = (-G_0 u) g_0$$

$$= (q_1 - p_1) G_1 + (q_2 - p_2) G_2 + (q_3 - p_3) G_3 = q - p.$$

Hence the direction of the line l is parallel to the line joining p to q.

iii. To find a point on the line l, observe that the plane u is orthogonal to the line l, since the direction of the normal vector to the plane u is parallel to the direction of the line l. Therefore, to find a point on the line l, we need only intersect the line l with the plane u. By Theorem 2.8, $u \wedge l$ is the intersection point of the line l and the plane u. Moreover by (*)

$$u \wedge l = u \wedge (g_0 v - G_0 u) = u \wedge (g_0 v) - u \wedge (G_0 u).$$

To analyze this expression, we shall consider each term in turn. For the second term, by Chapter I, Equation (3.13)

$$u \wedge (G_0 u) = \frac{u(G_0 u) + G_0 uu}{2} = \frac{G_0 uu + G_0 uu}{2} = (u \cdot u) G_0$$

For the first term, it follows from Chapter 1, Equations (3.13), (3.3), and (4.2) that

$$u \wedge (g_0 v) = \frac{u(g_0 v) + (g_0 v) u}{2} = g_0 \frac{vu - uv}{2}$$

$$= g_0 (v \wedge u) = g_0 G_0 (v \times u) = g_{0123} (v \times u).$$

Moreover,

$$v = v_q^\perp \times v_p^\perp = \left(\frac{v_p^\perp + v_q^\perp}{2} \right) \times (v_p^\perp - v_q^\perp)$$

$$u = (v_p^\perp - v_q^\perp),$$

so

$$v \times u = (v_q^\perp \times v_p^\perp) \times (v_p^\perp - v_q^\perp) = \left(\left(\frac{v_p^\perp + v_q^\perp}{2} \right) \times (v_p^\perp - v_q^\perp) \right) \times (v_p^\perp - v_q^\perp).$$

To simplify this last expression, we need to collapse a double cross product. To do so, we shall invoke the identity (see Chapter 1, Section 4.1, Exercise 2b)

$$(a \times b) \times b = (b \cdot a)b - (b \cdot b)a.$$

Applying this identity to the right-hand side of our formula for $v \times u$ yields

$$v \times u = \left((v_p^\perp - v_q^\perp) \cdot \left(\frac{v_p^\perp + v_q^\perp}{2} \right) \right)(v_p^\perp - v_q^\perp) - \left((v_p^\perp - v_q^\perp) \cdot (v_p^\perp - v_q^\perp) \right)\left(\frac{v_p^\perp + v_q^\perp}{2} \right).$$

But for vectors x of grade 3, $g_{0123}x^\perp = x$ (see Exercise 4). Hence

$$u \wedge (g_0 v) = g_{0123}(v \times u) = \left((v_q^\perp - v_p^\perp) \cdot \left(\frac{v_p^\perp + v_q^\perp}{2} \right) \right)(v_p - v_q) - (u \cdot u)\left(\frac{v_p + v_q}{2} \right).$$

Putting all our formulas together, we find that

$$u \wedge l = u \wedge (g_0 v) - u \wedge (G_0 u)$$

$$= \left((v_q^\perp - v_p^\perp) \cdot \left(\frac{v_p^\perp + v_q^\perp}{2} \right) \right)(v_p - v_q) - (u \cdot u)\left(\frac{v_p + v_q}{2} + G_0 \right).$$

Since

$$\frac{v_p + v_q}{2} + G_0 = \frac{(G_0 + v_p) + (G_0 + v_q)}{2} = \frac{p + q}{2}$$

$$v_p - v_q = p - q,$$

we arrive finally at the formula

$$u \wedge l \equiv \frac{p + q}{2} + \frac{\left((v_q^\perp - v_p^\perp) \cdot \left(\frac{v_p^\perp + v_q^\perp}{2} \right) \right)}{u \cdot u}(q - p).$$

Now $\dfrac{p+q}{2}$ is the midpoint of the line segment joining p to q, and $q-p$ is the vector from p to q. Therefore $u \wedge l$ is just the point $\dfrac{p+q}{2}$ displaced along the line from p to q, and hence is another point on the line joining p to q. $\Diamond$

Remark 2.11 (Plucker Coordinates)

The Plucker coordinates for the line l joining p to q are the pair

$$\left(q-p, (u \wedge l - G_0) \times (q-p)\right).$$

Moreover,

$$(u \wedge l - G_0) \times (q-p) = \left(\frac{p+q}{2} - G_0 + \frac{\left((v_q^{\perp} - v_p^{\perp}) \cdot \left(\dfrac{v_p^{\perp} + v_q^{\perp}}{2} \right) \right)}{u \cdot u} (q-p) \right) \times (q-p)$$

$$= \left(\frac{p+q}{2} - G_0 \right) \times (q-p)$$

Thus, as one might expect, the Plucker coordinates for the line l joining p to q are just

$$\left(q-p, \left(\frac{p+q}{2} - G_0 \right) \times (q-p) \right).$$

But wait a minute in the plane model of Clifford algebra, we cannot take the cross product of elements of grade 3; the cross product is defined only for elements of grade 1. Thus, to take the cross product in the plane model, we must work with the duals of these intuitive Plucker coordinates. Now notice that the pair (u, v), where

$$u = v_p^{\perp} - v_q^{\perp} = -(q-p)^{\perp}$$

$$v = v_q^{\perp} \times v_p^{\perp} = \left(\frac{v_p^{\perp} + v_q^{\perp}}{2} \right) \times (v_p^{\perp} - v_q^{\perp}) = -\left(\frac{p+q}{2} - G_0 \right)^{\perp} \times (q-p)^{\perp}$$

are, up to sign, precisely the duals that we need of the intuitive version of the Plucker coordinates. Therefore, in the plane model, in terms of Plucker coordinates

$$l = g_0 v - G_0 u = -g_0 \left(\left(\frac{p+q}{2} - G_0 \right)^{\perp} \times (q-p)^{\perp} \right) + G_0 (q-p)^{\perp}. \quad (2.18)$$

We close this section by showing how to apply Theorem 2.10 to develop a more direct method for determining whether or not a point p lies on a line l.

Corollary 2.12

Let l be a bivector representing a line. Then a point p lies on the line l if and only if $l^{\perp} \wedge p^{\perp} = 0$.

Proof: By Theorem 2.10 since the bivector l represents a line, $l = \left(p_2^{\perp} \wedge p_1^{\perp} \right)^{\perp}$ where p_1, p_2 are two points on the line l. Now p is a point on the line l if and only if there is a constant $\lambda \neq 0$ such that $p = (1-\lambda) p_1 + \lambda p_2$. Moreover,

$$p = (1-\lambda) p_1 + \lambda p_2 \Leftrightarrow p^{\perp} = (1-\lambda) p_1^{\perp} + \lambda p_2^{\perp} \Leftrightarrow p_2^{\perp} \wedge p_1^{\perp} \wedge p^{\perp} = 0$$

$$\Leftrightarrow l^{\perp} \wedge p^{\perp} = 0. \lozenge$$

Exercises

1. Show that the join $p \vee q$ is an antisymmetric, bilinear function. That is, show that

   a. $q \vee p = -p \vee q$

   b. $(ap + bq) \vee r = a(p \vee r) + b(q \vee r)$

   c. $p \vee (cq + dr) = c(p \vee q) + d(p \vee r)$

2. Show that, up to constant multiples, the line represented by the join of two points is independent of the choice of the two points on the line. (Hint: Apply Exercise 1).

3. Consider the planes u, v defined in the proof of Theorem 2.10. Show that $u \cdot v = 0$:

   a. by direct computation

   b. by using standard properties of the cross product.

4. Prove that for vectors x of grade 3, $g_{0123}x^{\perp} = x$.

5. Show that the bivector $p \vee v = \left(v^{\perp} \wedge p^{\perp}\right)^{\perp}$ represents the line through the point p parallel to the vector v.

6. Show that the bivector $l = \left(\left(q-p\right)^{\perp} \wedge p^{\perp}\right)^{\perp}$ represents the line through the point p parallel to the vector $q - p$.

7. Show that the duals of the Plucker coordinates and the dual Plucker coordinates for the same line are equal up to a constant multiple. (Hint: Compare Equations 2.8 and 2.18.)

8. Show that the point p lies on the line l if and only if the line $l^{\perp}$ lies on the plane $p^{\perp}$.

9. The bivector g_{21} represents the z-axis, and the trivector $G_0 + \lambda G_3$ represents points along the z-axis. Confirm Corollary 2.12 by verifying that $g_{21}^{\perp} \wedge (G_0 + \lambda G_3)^{\perp} = 0$.

---

3 Transformations: Rotors and Versors

In Part I we show how to compute translation, rotation, and reflection on vectors, points, and planes in the space of dual quaternions. In Sections 2.1 and 2.2, we show that multiplication by G_0 maps vectors, points, and planes in the space of dual quaternions into the corresponding vectors, points, and planes in the plane model of Clifford algebra. *Therefore, our approach to deriving the formulas for translation, rotation, and reflection in the plane model for dual quaternions will be to use the formulas for these transformations on vectors, points, and planes in the space of dual quaternions together with multiplication by G_0 to derive the corresponding formulas for translation, rotation, and reflection on vectors, points, and planes in the plane model of Clifford algebra.* We shall see that with this approach, the proofs of these transformation formulas simplify considerably; moreover, the formulas for these transformations on lines follow readily from the formulas for these transformations on planes, since the bivectors representing lines are wedge products of pairs of planes. To derive compact formulas for perspective and pseudo-perspective, we shall use much the same approach along with duality in the plane model of Clifford algebra.

3.1 Translation

We initiate our study of rigid motions with translations. To proceed, we need the following lemma.

Lemma 3.1

Let $T = 1 - \dfrac{1}{2} at$, where $t = t_1 g_{01} + t_2 g_{02} + t_3 g_{03}$ and a is a scalar. Then

a. $T^{-1} = 1 + \dfrac{1}{2} at$

b. $TG_0 = G_0 T^{-1}$.

Proof: Since $g_0^2 = 0$, it follows that $\left(1 - \dfrac{1}{2} at\right)\left(1 + \dfrac{1}{2} at\right) = 1$. Hence $T^{-1} = \left(1 + \dfrac{1}{2} at\right)$. Moreover, by Equations (1.3) and (1.4), $g_0 g_j G_0 = -G_0 g_0 g_j$; therefore $TG_0 = G_0 T^{-1}$. ◊

Theorem 3.2 (Translation)

Let $t = t_1 G_1 + t_2 G_2 + t_3 G_3$ be a unit vector and let $t = -G_0 t = t_1 g_{01} + t_2 g_{02} + t_3 g_{03}$ be the corresponding dual quaternion. Let a be a scalar, and set

$$T = 1 + \frac{1}{2} a G_0 t = 1 - \frac{1}{2} at.$$

Then the sandwiching map

$$x \mapsto TxT^{-1}$$

i. *leaves vectors unchanged*
ii. *translates points, lines, and planes by the distance a in the direction t.*

Proof: Since planes have grade 1, we begin with planes.

1. **Planes:** To establish this result for planes, we invoke the correspondence between planes in the plane model of Clifford algebra and planes in the algebra of dual quaternions. Consider then an arbitrary plane π, and let $\tilde{\pi} = -G_0 \pi$ be the dual quaternion corresponding to π. Then by Equation (2.2) $\pi = G_0 \tilde{\pi}$ and by Lemma (3.1) $TG_0 = G_0 T^{-1}$, so

$$T\pi T^{-1} = TG_0 \tilde{\pi} T^{-1} = G_0 T^{-1} \tilde{\pi} T^{-1}.$$

To understand the term $T^{-1}\tilde{\pi}T^{-1}$, observe that

$$T^{-1} = 1 + \frac{1}{2}at$$

$$t = -G_0t = t_1g_{01} + t_2g_{02} + t_3g_{03} = \left(t_1g_{32} + t_2g_{13} + t_3g_{21}\right)g_{0123} = \tilde{t}g_{0123}$$

where

$$\tilde{t} = t_1g_{32} + t_2g_{13} + t_3g_{21}$$

is a unit quaternion. Therefore

$$T^{-1} = 1 + \frac{1}{2}at = 1 + \frac{1}{2}a\tilde{t}g_{0123}$$

corresponds to a translation in the space of dual quaternions by the distance a in the direction t. In particular by Part I, Theorem 4.10, the sandwiching map

$$\tilde{\pi} \mapsto T^{-1}\tilde{\pi}T^{-1}$$

translates planes $\tilde{\pi}$ by the distance a in the direction t. Therefore, the map

$$\pi \mapsto T\pi T^{-1} = G_0(T^{-1}\tilde{\pi}T^{-1})$$

translates the corresponding planes $\pi = G_0\tilde{\pi}$ by the distance a in the corresponding direction $t = G_0t$.

2. **Lines:** Consider a line represented by a bivector l. Then there are two planes π_1, π_2 such that $l = \pi_2 \wedge \pi_1$. Therefore, by Chapter I, Equation (3.7)

$$TlT^{-1} = T\left(\pi_2 \wedge \pi_1\right)T^{-1} = \left(T\pi_2T^{-1}\right) \wedge \left(T\pi_1T^{-1}\right).$$

By part 1, the map $\pi \mapsto T\pi T^{-1}$ translates each of the planes π_1, π_2 by the distance a in the direction t. But translation is a rigid motion, so intersecting and then translating is equivalent to translating first and then intersecting. Therefore the map $l \mapsto TlT^{-1}$ translates the intersection line l by the distance a in the direction t.

3. **Points:** The proof for points is analogous to the proof for planes, this time invoking the correspondence between points in the plane model of Clifford algebra and points in the algebra of dual quaternions. Consider then an arbitrary point p, and let $\tilde{p} = -G_0 p$ be the dual quaternion corresponding to p. By Equation (2.4) $p = G_0 \tilde{p}$ and by Lemma (3.1) $TG_0 = G_0 T^{-1}$, so

$$TpT^{-1} = TG_0 \tilde{p} T^{-1} = G_0 T^{-1} \tilde{p} T^{-1}.$$

But again since

$$\tilde{t} = t_1 g_{32} + t_2 g_{13} + t_3 g_{21}$$

is a unit quaternion,

$$T^{-1} = 1 + \frac{1}{2} at = 1 + \frac{1}{2} a\tilde{t} g_{0123}$$

corresponds to a translation in the space of dual quaternions. In particular by Part I, Theorem 4.1, the map

$$\tilde{p} \mapsto T^{-1} \tilde{p} T^{-1}$$

translates points $\tilde{p}$ by the distance a in the direction t. Therefore the map

$$p \mapsto TpT^{-1} = G_0(T^{-1} \tilde{p} T^{-1})$$

translates the corresponding points $p = G_0 \tilde{p}$ by the distance a in the corresponding direction $t = G_0 t$. An alternative, more direct, computational proof is provided in Exercise 4.

4. **Vectors:** The proof for vectors mimics the proof for points, this time invoking the correspondence between vectors in the plane model of Clifford algebra and vectors in the algebra of dual quaternions—see Exercise 1. An alternative, more direct, computational proof is provided in Exercise 2. ◊

Notice how the plane model reverses the signs of T from dual quaternions and maps one factor of T to T^{-1}. Notice too that for planes if the direction of translation is perpendicular to the normal to the plane, then the points on the plane are translated along the plane, but the plane itself—that is, the vector representing the plane—does not change. Similarly for lines, if the direction of translation is parallel to the line, then the points on the line are translated along the line, but the line itself—that is, the bivector representing the line—does not change (see Exercise 9). For similar observations for planes in the space of dual quaternions, see Part I, Section 4.4.

The expression $T = 1 - \frac{1}{2}at$ in Theorem 3.2 that represents translation is called a *rotor* because translation is a rotation in 8-dimensions (see Part I, Section 5). The translation rotor can also be expressed as an exponential, since $g_0^2 = 0$ implies that $T = e^{-\frac{1}{2}at}$ (see Exercise 10).

Exercises

1. Prove Theorem 3.2 for vectors by invoking the correspondence between vectors in the plane model of Clifford algebra and vectors in the algebra of dual quaternions.

2. Prove that Theorem 3.2 for vectors v by direct computation using the identity:

$$v = v_1 G_1 + v_2 G_2 + v_3 G_3 = g_0 \left(v_1 g_{32} + v_2 g_{13} + v_3 g_{21} \right) = g_0 v.$$

3. Prove Theorem 3.2 for the origin G_0 by direct computation using the identities:

$$T G_0 T^{-1} = T^2 G_0 \text{ and } G_0 t = -t G_0.$$

4. Prove Theorem 3.2 for points by direct computation using Exercises 2 and 3.

5. Prove Theorem 3.2 for points and vectors using the fact that for any point p or vector v, there are 3 planes π_1, π_2, π_3 that intersect in the point p or the vector v.

6. Prove that Theorem 3.2 is valid for planes if and only if Theorem 3.2 is valid for points by using the fact that a point p lies on a plane π if and only if $\pi \wedge p = 0$.

7. Prove that the plane at infinity is unaffected by the translation map in Theorem 3.2—that is, prove that $T g_0 T^{-1} = g_0$.

8. Consider a plane π with unit normal vector v. Let $u = v g_{0123} = v^{\perp}$ be a vector parallel to v, and set $T = 1 + \frac{1}{2} a G_0 u$.

   a. Show that $T \pi T^{-1} = \pi - a g_0$.

b.   Give a geometric reason for the result in part a.

9. Consider a bivector $l = l_{32}g_{32} + l_{13}g_{13} + l_{21}g_{21} + l_{01}g_{01} + l_{02}g_{02} + l_{03}g_{03}$ representing a line. Let $\mathbf{u} = lg_0$, and set $T = 1 + \dfrac{1}{2}G_0\mathbf{u}$

     a.   Show that $TlT^{-1} = l$ in two ways:

         i.   By appealing to Theorem 3.2.

         ii.   By direct computation.

     b.   Give a geometric reason for the result in part a.

10. Let $T = 1 - \dfrac{1}{2}at$ be the translation rotor in Theorem 3.2. Show that $T = e^{-\frac{1}{2}at}$. (Hint: Use the Taylor expansion for e^x)

3.2 Rotation

We continue our study of rigid motions with rotations. Translations depend on the bivectors g_{01}, g_{02}, g_{03}; rotations depend on the complementary bivectors g_{32}, g_{13}, g_{21}—that is, on the quaternions. To proceed, we need the following analogue of Lemma 3.1.

Lemma 3.3

Let $R = \cos(\theta/2) + \tilde{v}\sin(\theta/2)$, where $\tilde{v} = v_1 g_{32} + v_2 g_{13} + v_3 g_{21}$ is a unit quaternion. Then

     a. $R^{-1} = R^*$

     b. $RG_0 = G_0R.$

Proof: Since $\tilde{v}$ is a unit quaternion, $\tilde{v}^2 = -1$. Hence

$$\left(\cos(\theta/2) + \tilde{v}\sin(\theta/2)\right)\left(\cos(\theta/2) - \tilde{v}\sin(\theta/2)\right) = 1.$$

Therefore $R^{-1} = \cos(\theta/2) - \tilde{v}\sin(\theta/2) = R^*$

Moreover, by Equation (1.4), $\tilde{v}G_0 = G_0\tilde{v}$; therefore $RG_0 = G_0R$. $\Diamond$

Theorem 3.4 (Rotation about a Line through the Origin)

Let $\tilde{v} = v_1 g_{32} + v_2 g_{13} + v_3 g_{21}$ be a unit quaternion representing a line pass-ing through the origin, and let

$$R = \cos(\theta/2) + \tilde{v}\sin(\theta/2).$$

Then the sandwiching map

$$x \mapsto RxR^{-1}$$

rotates vectors, points, lines, and planes by the angle θ around the line $\tilde{v}$.

Proof: Once again we begin with planes.

1. **Planes**: To establish this result for planes, we again invoke the correspondence between planes in the plane model of Clifford algebra and planes in the algebra of dual quaternions. Consider then an arbitrary plane

$$\pi = ag_1 + bg_2 + cg_3 + dg_0$$

in the plane model of Clifford algebra, and let

$$\tilde{\pi} = -G_0\pi = ag_{32} + bg_{13} + cg_{21} + dg_{0123}$$

be the dual quaternion corresponding to π. Then by Equation (2.2) $\pi = G_0\tilde{\pi}$ and by Lemma (3.3) $R^{-1} = R^*$ and $RG_0 = G_0R$, so

$$R\pi R^{-1} = RG_0\tilde{\pi}R^{-1} = G_0R\tilde{\pi}R^*.$$

But by Part I, Theorem 4.10, since R corresponds to a unit quater-nion in the space of dual quaternions, the map

$$\tilde{\pi} \mapsto R\tilde{\pi}R^*$$

rotates the plane $\tilde{\pi}$ by the angle θ around the line through the origin parallel to the unit vector $\tilde{v}\varepsilon$. Therefore

$$\pi \mapsto R\pi R^{-1} = G_0\left(R\tilde{\pi}R^*\right)$$

rotates the corresponding plane π by the angle θ around the line through the origin parallel to the corresponding unit vector $G_0\tilde{v}g_{0123} = \tilde{v}g_0$ which by Theorem 3.6 is the line $\tilde{v}$.

2. **Lines:** Let the bivector l represent a line in the plane model of Clifford algebra. Then there are planes π_1, π_2 such that the line l is the intersection of the planes π_1, π_2; hence $l = \pi_2 \wedge \pi_1$. Therefore by Chapter 1, Equation (3.7)

$$RlR^{-1} = R(\pi_2 \wedge \pi_1)R^{-1} = \left(R\pi_2 R^{-1}\right) \wedge (R\pi_1 R^{-1}).$$

Thus, since by part 1 $\pi_1 \mapsto R\pi_1 R^{-1}$ and $\pi_2 \mapsto R\pi_2 R^{-1}$ rotate the planes π_1, π_2 by the angle θ around the line $\tilde{v}$ passing through the origin, the map $l \mapsto RlR^{-1}$ must also rotate the intersection line l by the angle θ around the line $\tilde{v}$ passing through the origin.

3. **Points and Vectors:** The proofs for points and vectors are analogous to the proof for planes, this time invoking the correspondence between points and vectors in the plane model of Clifford algebra and points and vectors in the algebra of dual quaternions—see Exercise 1. Alternative proofs are provided in Exercise 2. $\lozenge$

Rigid motions can now be computed by composing rotations and translations using the Clifford product:

Rotation followed by Translation : $x \mapsto T\left(RxR^{-1}\right)T^{-1} = (TR)x(TR)^{-1}$

Translation followed by Rotation : $x \mapsto R\left(TxT^{-1}\right)R^{-1} = (RT)x(RT)^{-1}.$

Thus, the composite of a rotation and a translation can be represented by the Clifford product of the corresponding transformations. Notice the similarity to composing these transformations using dual quaternions in Part I, Section 4.1, but here conjugates are replaced by inverses.

The expression $R = \cos(\theta/2) + \tilde{v}\sin(\theta/2)$ in Theorem 3.4 that represents rotation is also called a *rotor*. The rotation rotor too can be expressed as an exponential: $R = e^{\tilde{v}\theta/2}$ (see Exercise 7).

Theorem 3.5 (Rotation around an Arbitrary Line)

Consider an arbitrary line represented by a bivector $l = l_{32}g_{32} + l_{13}g_{13} + l_{21}g_{21} + l_{01}g_{01} + l_{02}g_{02} + l_{03}g_{03}$ with unit quaternion $\tilde{v} = l_{32}g_{32} + l_{13}g_{13} + l_{21}g_{21}$ representing a line through the origin parallel to the line l. Let $C = G_0 + t_1 G_1 + t_2 G_2 + t_3 G_3$ be a point on the line l, and set

$$\tilde{t} = t_1 g_{32} + t_2 g_{13} + t_3 g_{21} = G_0(C - G_0)^\perp$$

$$R = \cos(\theta/2) + \tilde{v}\sin(\theta/2)$$

$$S = R + \sin(\theta/2)\left(\tilde{v} \times_q \tilde{t}\right)g_{0123}$$

(Here $\times_q$ denotes the quaternion cross product — Chapter 1, Section 4.1)

Then the map $x \mapsto SxS^{-1}$ rotates vectors, points, lines, and planes by the angle θ about the line l.

Proof: Let

$$\mathbf{t} = C - G_0 = t_1 G_1 + t_2 G_2 + t_3 G_3.$$

$$-G_0\mathbf{t} = t_1 g_{01} + t_2 g_{02} + t_3 g_{03} = \left(t_1 g_{32} + t_2 g_{13} + t_3 g_{21}\right)g_{0123} = \tilde{t}g_{0123}.$$

Set

$$T = 1 + \frac{1}{2}G_0\mathbf{t} = 1 - \frac{1}{2}\tilde{t}g_{0123}.$$

Then by Theorem 3.2, T represents translation by the vector $\mathbf{t}$, which maps the origin G_0 to the point C. Hence T^{-1} represents translation by the vector $-\mathbf{t}$, which maps the point C to the origin G_0. Since by Theorem 3.4 the map $x \mapsto RxR^{-1}$ rotates vectors, points, lines, and planes by the angle θ about the line $\tilde{v}$ passing through the origin parallel to the line l, it remains only to show that $S = TRT^{-1}$. Now the rest of this proof is almost step by step identical to the similar steps in the proof of Theorem 4.2 in Part I (see Exercise 8). Notice, in particular, that $\tilde{v}$ and $\tilde{t}$ are pure quaternions, so the quaternion cross product

$\tilde{v} \times_q \tilde{t}$ makes sense in the space of quaternions and is a bivector in the Clifford algebra for dual quaternions (see Chapter 1, Theorem 4.1). ◊

Theorem 3.6 (Screw Transformations— Simultaneous Rotation and Translation)

Consider a bivector $l = l_{32}g_{32} + l_{13}g_{13} + l_{21}g_{21} + l_{01}g_{01} + l_{02}g_{02} + l_{03}g_{03}$ representing an arbitrary line with unit quaternion $\tilde{v} = l_{32}g_{32} + l_{13}g_{13} + l_{21}g_{21}$ representing a line through the origin parallel to the line l. Let $C = G_0 + t_1G_1 + t_2G_2 + t_3G_3$ be a point on the line l, and set

$$\tilde{t} = t_1 g_{32} + t_2 g_{13} + t_3 g_{21} = G_0(C - G_0)^{\perp}$$

$$R = \cos(\theta / 2) + \tilde{v} \sin(\theta / 2)$$

$$S = R + \sin(\theta / 2)\left(\tilde{v} \times_q \tilde{t}\right)g_{0123}$$

$$M = S + \frac{1}{2}d\left(\sin(\theta / 2) - \tilde{v} \cos(\theta / 2)\right)g_{0123}$$

Then the map $x \mapsto MxM^{-1}$ first rotates vectors, points, lines, and planes by the angle θ about the line l and then translate the resulting object in the direction lg_0 of the line l by the distance d.

Proof: By Theorem 3.5 the map $x \mapsto SxS^{-1}$ represents rotation by the angle θ about the line l. Let

$$\mathbf{v} = lg_0 = \tilde{v}g_0$$

$$T = 1 + \frac{d}{2}G_0\mathbf{v}.$$

By Theorem 3.2, the map $x \mapsto TxT^{-1}$ represents translation by the distance d in the direction of the vector $\mathbf{v} = lg_0$, which is a vector parallel to the direction of the line l. Therefore, we need only show that $M = TS$. But

$$G_0 \mathbf{v} = G_0 \left(\tilde{v} g_0 \right) = -\tilde{v} g_{0123}$$

$$T = 1 + \frac{d}{2} G_0 \mathbf{v} = 1 - \frac{d}{2} \tilde{v} g_{0123}.$$

Now we can simply mimic the proof in Part I, Theorem 4.4:

$$TS = (1 - \frac{1}{2} d\tilde{v} g_{0123})(R + \sin(\theta/2)\left(\tilde{v} \times_q \tilde{t}\right) g_{0123}) = S - \frac{1}{2} d\tilde{v} R g_{0123}.$$

Moreover since $\tilde{v}$ is a unit quaternion $\tilde{v}^2 = -1$, so

$$\tilde{v}R = \tilde{v}\left(\cos(\theta/2) + \tilde{v}\sin(\theta/2)\right) = -\sin(\theta/2) + \tilde{v}\cos(\theta/2).$$

Hence

$$TS = S + \frac{1}{2} d\left(\sin(\theta/2) - \tilde{v}\cos(\theta/2)\right) g_{0123} = M. \Diamond$$

Remark 3.7

In Theorems 3.5 and 3.6, $\tilde{v}$ is a unit quaternion. Therefore, by Section 2.3.1 we can choose the point C on the line l and the vector $\mathbf{t}$ by setting

$$C = G_0 - \tilde{u} \wedge v$$

$$\mathbf{t} = C - G_0 = -\tilde{u} \wedge v$$

where

$$v = \tilde{v} G_0 = l_{32} g_1 + l_{13} g_2 + l_{21} g_3$$

$$\tilde{u} = -\left(l_{01} g_{01} + l_{02} g_{02} + l_{03} g_{03}\right).$$

Thus, explicitly we can set

$$\mathbf{t} = \underbrace{\left(l_{13}l_{03} - l_{21}l_{02}\right)}_{t_1} G_1 + \underbrace{\left(l_{21}l_{01} - l_{32}l_{03}\right)}_{t_2} G_2 + \underbrace{\left(l_{32}l_{02} - l_{13}l_{01}\right)}_{t_3} G_3$$

$$\tilde{t} = \underbrace{\left(l_{13}l_{03} - l_{21}l_{02}\right)}_{\tilde{t}_1} g_{32} + \underbrace{\left(l_{21}l_{01} - l_{32}l_{03}\right)}_{\tilde{t}_2} g_{13} + \underbrace{\left(l_{32}l_{02} - l_{13}l_{01}\right)}_{\tilde{t}_3} g_{21}.$$

Exercises

1. Prove Theorem 3.4 for points and vectors by invoking the correspondence between points and vectors in the plane model of Clifford algebra and points and vectors in the algebra of dual quaternions.

2. Prove Theorem 3.4 for points and vectors using the fact that for any point p or vector v, there are 3 planes π_1, π_2, π_3 that intersect in the point p or the vector v.

3. Prove that Theorem 3.4 is valid for planes if and only if Theorem 3.4 is valid for points by using the fact that a point p lies on a plane π if and only if $\pi \wedge p = 0$.

4. Prove that the origin is unaffected by the rotation map in Theorem 3.4—that is, prove that $RG_0R^{-1} = G_0$.

5. Prove that Theorem 3.4 is valid for points and if and only if Theorem 3.4 is valid for vectors. (Hint: Use Exercise 4.)

6. Prove that the plane at infinity is unaffected by the rotation map in Theorem 3.4—that is, prove that $Rg_0R^{-1} = g_0$.

7. Let $R = \cos(\theta/2) + \tilde{v}\sin(\theta/2)$ be the rotation rotor in Theorem 3.4. Show that $R = e^{\tilde{v}\theta/2}$. (Hint: Use the Taylor expansion for e^x.)

8. Complete the proof of Theorem 3.5 by mimicking the proof of Theorem 4.2 in Part I.

3.3 Reflection

Reflections in the plane model are easier than both translations and rotations.

Theorem 3.8 (Reflection in an Arbitrary Plane)

Let π represent an arbitrary plane with a unit normal vector. Then the sandwiching map

$$x \mapsto \pi x \pi$$

reflects vectors, points, lines, and planes in the plane π.

Proof: First observe that since π is a plane, there is a dual quaternion $\tilde{\pi}$ such that $\pi = G_0\tilde{\pi}$.

1. **Planes**: To establish this result for planes, we shall once again invoke the correspondence between planes in the plane model of Clifford algebra and planes in the algebra of dual quaternions. Consider then an arbitrary plane τ, and let $\tilde{\tau} = -G_0\tau$ be the dual quaternion corresponding to τ. Then by Equation (2.6) $\tau G_0 = \tilde{\tau}^*$, so

$$\pi\tau\pi = (G_0\tilde{\pi})(\tau)(G_0\tilde{\pi}) = G_0\left(\tilde{\pi}\tilde{\tau}^*\tilde{\pi}\right).$$

Moreover, by Part I Theorem 4.20, the map

$$\tilde{\tau} \mapsto -\tilde{\pi}\tilde{\tau}^*\tilde{\pi}$$

reflects the plane $\tilde{\tau}$ in the plane $\tilde{\pi}$. Therefore the map

$$\tau \mapsto -\pi\tau\pi = G_0\left(-\tilde{\pi}\tilde{\tau}^*\tilde{\pi}\right)$$

reflects the corresponding plane $\tau = G_0\tilde{\tau}$ in the corresponding plane $\pi = G_0\tilde{\pi}$. But the expressions $-\pi\tau\pi$ and $\pi\tau\pi$ represent the same plane. Therefore, the map $\tau \mapsto \pi\tau\pi$ reflects planes τ in the plane π.

2. **Lines**: Let the bivector l represent a line in the plane model of Clifford algebra. Then there are planes π_1, π_2 such that the line l is the intersection of the planes π_1, π_2; hence $l = \pi_2 \wedge \pi_1$. Therefore by Chapter I, Equation (3.7)

$$\pi l \pi^{-1} = \pi(\pi_2 \wedge \pi_1)\pi^{-1} = \left(\pi\pi_2\pi^{-1}\right) \wedge (\pi\pi_1\pi^{-1}). \qquad (*)$$

Moreover, since π has grade 1 with a unit normal vector, $\pi^2 = \pi \cdot \pi = 1$, so $\pi^{-1} = \pi$. Substituting this formula for π^{-1} into (*) yields

$$\pi l \pi = \left(\pi\pi_2\pi\right) \wedge (\pi\pi_1\pi).$$

Thus, since by part 1 the map $\tau \mapsto \pi\tau\pi$ reflects the planes π_1, π_2 in the plane π, the map $l \mapsto \pi l \pi$ must also reflect the intersection line l in the plane π.

3. **Points and Vectors**: The proof for points and vectors is analogous to the proof for planes, this time invoking the correspondence between points and vectors in the plane model of Clifford algebra and points and vectors in the algebra of dual quaternions and using as well the identity $-pG_0 = \tilde{p}^*$—see Exercise 1. Alternative proofs are provided in Exercises 2 and 3. ◊

Remark 3.9

Notice how the ugly conjugates π^* and p^* that appear in the dual quaternion sandwiching formulas for reflection in Part I, Theorem 4.20 disappear from these elegant Clifford algebra sandwiching formulas for reflection. Also, the formula for reflection on lines is now the same as the formula or reflection on points, vectors, and planes.

The plane π in Theorem 3.8 that represents reflection is called a *versor*. Unlike rotors, versors cannot be expressed as exponentials.

Exercises

1. Prove Theorem 3.8 for points and vectors by invoking the correspondence between points and vectors in the plane model of Clifford algebra and points and vectors in the algebra of dual quaternions and applying the identity $-pG_0 = \tilde{p}^*$ from Equation (2.6).

2. Prove that Theorem 3.8 is valid for planes if and only if Theorem 3.8 is valid for points by using the fact that a point p lies on a plane τ if and only if $\tau \wedge p = 0$.

3. Prove Theorem 3.8 for vectors using the duality between vectors and normal vectors or equivalently between vectors and planes through the origin.

4. Let π_1, π_2 be parallel planes with unit normal vectors. Show that:

   a. $\pi_1\pi_2$ represents a translation rotor.

   b. Every translation is the product of two reflections in parallel planes.

3.4 Perspective and Pseudo-Perspective

Here we shall invoke the translation rotor along with duality to compute perspective and pseudo-perspective in the plane model for dual quaternions. We begin with the following lemma relating duality in the space of dual quaternions with duality in the plane model.

Lemma 3.10

Let x be a mass-point or a vector, and let $\tilde{x} = -G_0 x$ be the dual quaternion corresponding to x. Then $\tilde{x}^{\#} = G_0 x^{\perp}$.

Proof: We shall prove this result for mass-points; the proof for vectors is similar, just set the mass to zero. Consider then a mass-point p and the corresponding dual quaternion $\tilde{p}$:

$$p = p_0 G_0 + p_1 G_1 + p_2 G_2 + p_3 G_3 = G_0 \tilde{p}$$

$$\tilde{p} = -G_0 p = p_0 + p_1 g_{01} + p_2 g_{02} + p_3 g_{03} = p_0 + (p_1 g_{32} + p_2 g_{13} + p_3 g_{21}) g_{0123}.$$

Now by definition

$$p^{\perp} = -\left(p_0 g_0 + p_1 g_1 + p_2 g_2 + p_3 g_3\right)$$

$$\tilde{p}^{\#} = p_0 g_{0123} + p_1 g_{32} + p_2 g_{13} + p_3 g_{21}$$

$$= -G_0 (p_0 g_0 + + p_1 g_1 + p_2 g_2 + p_3 g_3) = G_0 p^{\perp}. \ \Diamond$$

Theorem 3.11 (Perspective Projection from the Origin)

Let

- $E = G_0 = $ *eye point*
- $t = t_1 G_1 + t_2 G_2 + t_3 G_3 = g_0 \left(t_1 g_{32} + t_2 g_{13} + t_3 g_{21}\right) = g_0 t' = $ *a unit vector*
- $t = -G_0 t = g_0 \left(t_1 g_1 + t_2 g_2 + t_3 g_3\right) = -g_0 t^{\perp}$
- $\pi = g_0 + \dfrac{1}{d} t^{\perp} = $ *the perspective plane at a distance d from the eye point $E = G_0$ perpendicular to the unit normal vector $t^{\perp}$*
- $T = 1 + \dfrac{1}{2d} t = $ *a translation rotor*

Then $\left(T(P - E)^{\perp} T^{-1}\right)^{\perp}$ is a mass-point (mP', m) where:

- *the point P' is located at the perspective projection of the point P from the eye point $E = G_0$ onto the plane π;*
- *the mass $m = s / d$ is equal to the distance s of the point P from the plane through the eye point $E = G_0$ perpendicular to the unit normal vector $t^{\perp}$ divided by the fixed distance d from the eye point E to the perspective plane π.*

Proof: By Theorem 7.2 in Part I, in the space of dual quaternions the mass-point representing perspective projection into the plane π is given by the dual quaternion $\left(T^{-1}\left(\widetilde{P-E}\right)^{\#}T^{-1}\right)^{\#}$, where $\widetilde{P-E} = -G_0(P-E)$ is the dual quaternion corresponding to the vector $P-E$, and $T^{-1} = 1 - \dfrac{1}{2d}t = 1 - \dfrac{1}{2d}t'g_{0123}$ is a translation operator in the space of dual quaternions. To convert from dual quaternions to Clifford algebra, we multiply the dual quaternion terms by G_0. Therefore in Clifford algebra this dual quaternion formula for perspective projection becomes

$$\left(T^{-1}(\widetilde{P-E})^{\#}T^{-1}\right)^{\#} \rightarrow G_0\left(G_0\left(T^{-1}(\widetilde{P-E})^{\#}T^{-1}\right)\right)^{\#}.$$

Now by Lemma 3.10, $-G_0\tilde{p}^{\#} = p^{\perp}$, and by Lemma 3.1, $G_0T^{-1} = TG_0$. Hence

$$G_0(G_0(T^{-1}(\widetilde{P-E})^{\#}T^{-1}))^{\#} = -G_0(T(-G_0)(\widetilde{P-E})^{\#}T^{-1})^{\#} = (T(P-E)^{\perp}T^{-1})^{\perp}.$$

For a more direct, computational proof, see Exercise 1. ◊

Our next result shows that the rotation rotor commutes with duality. Therefore, we shall see that we can compose rotation with perspective and pseudo-perspective in the same way that we can compose rotation with translation: by multiplying the two transformations.

Lemma 3.12

Let

$$\tilde{u} = u_1 g_{32} + u_2 g_{13} + u_3 g_{21} = unit \ quaternion \ representing \ a \ line \ passing$$
$$through \ the \ origin$$

$$R = \cos(\theta/2) + \tilde{u}\sin(\theta/2) = a \ rotation \ rotor$$

$$v = v_1 G_1 + v_2 G_2 + v_3 G_3 = an \ arbitrary \ vector$$

Then

$$\left(RvR^{-1}\right)^{\perp} = Rv^{\perp}R^{-1}.$$

Thus, rotating vectors around a line through the origin commutes with duality.

Proof: Observe that

$$v = v_1G_1 + v_2G_2 + v_3G_3 = \left(v_1g_1 + v_2g_2 + v_3g_3\right)g_{0123} = -v^{\perp}g_{0123}.$$

Therefore, since $g_{0123}R^{-1} = R^{-1}g_{0123}$, it follows that

$$RvR^{-1} = R\left(-v^{\perp}g_{0123}\right)R^{-1} = \left(-Rv^{\perp}R^{-1}\right)g_{0123} = \left(-Rv^{\perp}R^{-1}\right)^{\perp}$$

Now recall that by Equation (2.16) if grade x is odd, then $\left(x^{\perp}\right)^{\perp} = -x$. Therefore taking the dual of both sides of the previous formula yields

$$\left(RvR^{-1}\right)^{\perp} = Rv^{\perp}R^{-1}. \diamond$$

Since on vectors rotation and duality commute, we can compose rotation and perspective projection in the same way that we compose rotation and translation: by taking the product of their representations. Indeed, by Lemma 3.12

Rotation followed by Perspective

$$P \mapsto \left(T\left(R(P-E)R^{-1}\right)^{\perp} T^{-1}\right)^{\perp} = \left((TR)(P-E)^{\perp}(TR)^{-1}\right)^{\perp}$$

Perspective followed by Rotation

$$P \mapsto R\left(T(P-E)^{\perp} T^{-1}\right)^{\perp} R^{-1} = \left((RT)(P-E)^{\perp}(RT)^{-1}\right)^{\perp}$$

Note that here the rotation rotor R and the perspective map T are both with respect to the eye point at the origin—that is, the rotation is around a line through the origin.

Theorem 3.13 (Pseudo-Perspective)

Let

- $E = G_0 - dt = $ *eye point*
- $t = t_1G_1 + t_2G_2 + t_3G_3 = g_0\left(t_1g_{32} + t_2g_{13} + t_3g_{21}\right) = g_0t' = $ *a unit vector*
- $t = -G_0t = g_0\left(t_1g_1 + t_2g_2 + t_3g_3\right) = -g_0t^{\perp}$
- $T = 1 + \dfrac{1}{2d}t = $ *a translation rotor.*

Then the map $P \rightarrow \left(TP^{\perp}T^{-1}\right)^{\perp}$ *transforms the eye point E to the point at infinity in the direction* t *and transforms a viewing frustum to a rectangular box.*

Proof: By Theorem 7.3 in Part I, in the space of dual quaternions pseudo-perspective is given by the map $\tilde{P} \to (T^{-1}\tilde{P}^{\#}T^{-1})^{\#}$, where $\tilde{P} = -G_0 P$ and $T^{-1} = 1 - \dfrac{1}{2d}t = 1 - \dfrac{1}{2d}t' g_{0123}$ is a translation operator in the space of dual quaternions. Again, to convert from dual quaternions to Clifford algebra, we multiply the dual quaternion terms by G_0. Therefore in Clifford algebra this dual quaternion formula for pseudo-perspective becomes

$$P \to G_0 \left(G_0 (T^{-1}\tilde{P}^{\#}T^{-1}) \right)^{\#}.$$

Now by Lemma 3.10, $p^{\perp} = -G_0\tilde{p}^{\#}$, and by Lemma 3.1, $G_0T^{-1} = TG_0$. Therefore

$$P \to G_0 \left(G_0 (T^{-1}\tilde{P}^{\#}T^{-1}) \right)^{\#} = -G_0 \left(T(-G_0)\tilde{P}^{\#}T^{-1} \right)^{\#} = \left(TP^{\perp}T^{-1} \right)^{\perp}.$$

For a more direct, computational proof, see Exercise 2. ◊

In our theorems for perspective and pseudo-perspective, we have located the eye point E at a special canonical position for which the result is especially easy to compute. To deal with arbitrary eye points E', we can simply translate the entire scene—points and planes—by the vector $v = E - E'$ before performing perspective or pseudo-perspective and then after performing perspective or pseudo-perspective with the eye point in the canonical position E, translate the result by $-v = E' - E$ to move the result back to the required position with the eye at E'. For perspective, the vector $P - E$ is unaffected by translation, so we need only perform the final translation and can dispense altogether with the initial translation.

Exercises

1. Here we provide a more direct, computational proof of Theorem 3.11. Let $P = G_0 + v_P$. Using the notation of Theorem 3.11, show that:

   a. $T(P-E)^{\perp}T^{-1} = v_p^{\perp} - \dfrac{v_p^{\perp} \cdot t^{\perp}}{d} g_0$

   (Hint: First show that $\dfrac{tv_p^{\perp} - v_p^{\perp}t}{2} = -\left(v_p^{\perp} \cdot t^{\perp} \right) g_0$.)

   b. $\left(T(P-E)^{\perp}T^{-1} \right)^{\perp} \equiv G_0 + \left(\dfrac{d}{v_P \cdot t} \right) v_P$

   c. $\left(T(P-E)^{\perp}T^{-1} \right)^{\perp} \wedge \pi = \left(\left(v_P^{\perp} \cdot t^{\perp} \right) g_{0123} - v_P \wedge t^{\perp} \right)$

   d. $\left(v_P^{\perp} \cdot t^{\perp} \right) g_{0123} = v_P \wedge t^{\perp}$

Conclude that $\left(T(P-E)^{\perp}T^{-1}\right)^{\perp}$ is a mass-point (mP',m) where:

e. the point P' is located at the perspective projection of the point P from the eye point $E = G_0$ onto the plane π;

f. the mass $m = s\,/\,d$ is equal to the distance s of the point P from the plane through the eye point $E = G_0$ perpendicular to the unit normal vector $t^{\perp}$ divided by the fixed distance d from the eye point E to the perspective plane π.

2. Give a direct, computational proof of Theorem 3.13. In particular, using the notation of Theorem 3.13, show that:

a. $\left(TE^{\perp}T^{-1}\right)^{\perp} = dt$

b. $\left(Tdt^{\perp}T^{-1}\right)^{\perp} = -(G_0 + dt)$

c. $\left(Tt_{*}^{\perp}T^{-1}\right)^{\perp} \equiv -t_{*}$ where t_{*} is any vector perpendicular to t.

Conclude that the map $P \to \left(TP^{\perp}T^{-1}\right)^{\perp}$ transforms the eye point E to the point at infinity in the direction t and transforms a viewing frustum to a rectangular box.

3. Let P be an arbitrary point, and suppose that the eye point $E = G_0$.

a. Show that $l = \left(\left(P-E\right)^{\perp} \wedge E^{\perp}\right)^{\perp}$ is a bivector representing the line through the eye point E parallel to the vector $P-E$. (Hint: See Section 2.4.3, Exercise 6.)

b. Conclude that $\pi \wedge l = \pi \wedge \left(\left(P-E\right)^{\perp} \wedge E^{\perp}\right)^{\perp}$ is the perspective projection of the point P onto the plane π.

c. Generalize parts a and b to the case where the eye is placed at an arbitrary location.

4. Let π represent a plane through the origin, and let v be an arbitrary vector (grade 3).

a. Show that $\left(\pi v\pi\right)^{\perp} = -\pi v^{\perp}\pi$.

b. Conclude that reflecting a vector in a plane passing through the origin commutes with duality.

5. In this exercise we shall show how to compute perspective projection using the rotation rotor along with a novel form of duality. Define a new duality operator $x^{\&}$ by setting

- $x^{\&} = x^{\perp}\ grade(x)$ is even
- $g_j^{\&} = G_j$ and $G_j^{\&} = g_j\ j = 1,2,3$.
- $g_0^{\&} = g_0$ and $G_0^{\&} = G_0$

and extending to arbitrary multivectors by linearity. Notice that $\left(x^{\&}\right)^{\&} = x$ for all x.

Now to perform perspective projection using the rotation rotor, let

- $u = $ a unit quaternion
- $v = g_0 u = $ a unit vector
- $E = G_0 - v = $ eye point
- $\pi = v^{\perp} = $ perspective plane through the origin perpendicular to the unit vector v
- $R = \cos(\pi/4) - \sin(\pi/4)u = $ rotation rotor representing rotation by $-\pi/2$ around line through the origin parallel to the unit vector v

Show that $\left(R(P-E)^{\&} R\right)^{\&}$ is a mass-point (mP', m), where:

- the point P' is located at the perspective projection of the point P from the eye point E onto the plane π;
- the mass m is equal to the distance d of the point P from the plane through the eye point E perpendicular to the unit vector v.

(Hint: Recall that $g_i = G_0 g_{jk}$ for ijk an even permutation of 321, and apply [9, Chapter 6, Theorem 6.8]. Compare to Part I, Section 1.7.3, Exercise 5.)

6. Repeat Exercise 5 this time with

- $u = $ a unit quaternion
- $v = g_0 u = $ a unit vector
- $E = G_0 + \left(\cot(\theta) - \csc(\theta)\right)v = $ eye point
- $\pi = \cot(\theta)g_0 + v^{\perp} = $ perspective plane perpendicular to the unit vector v at a distance $\csc(\theta)$ from the eye point E
- $R = \cos(\theta/2) - \sin(\theta/2)u = $ rotation rotor representing rotation by $-\theta$ around the line through the origin parallel to the unit vector v

Show that $\left(R(P-E)^{\&} R\right)^{\&}$ is a mass-point (mP', m), where:

- the point P' is located at the perspective projection of the point P from the eye point E onto the plane π;
- the mass $m = d\sin(\theta)$, where d is the distance of the point P from the plane through the eye point E perpendicular to the unit vector v.

(Hint: Recall that $g_i = G_0 g_{jk}$ for ijk an even permutation of 321, and apply [9, Chapter 6, Theorem 6.9]. Compare to Part I, Section 1.7.3, Exercise 6.)

4 Insights

Here we list five of the main insights that guided our intuition for investigating and deriving the properties of the plane model of Clifford algebra for dual quaternions.

1. Quaternions and dual quaternions both reside inside the plane model
   - $g_{32} \leftrightarrow i \quad g_{13} \leftrightarrow j \quad g_{21} \leftrightarrow k \quad g_{0123} \leftrightarrow \varepsilon$

2. The origin G_0 maps from dual quaternion to Clifford algebra representations of geometry
   - $x \mapsto G_0 x$

3. Separation of operators and operands
   - Operators—Translation and rotation rotors have even grade
   - Operands—Points and planes have odd grade
4. Planes are the fundamental geometric objects
   - Lines are intersections of two planes
   - Points and vectors are intersections of three planes
5. Duality has four important applications
   - Cross products
   - Distance formulas
   - Joins of two points
   - Perspective and pseudo-perspective

5 Formulas

We provide a summary here for easy future reference of the main formulas we have encountered for the plane model of dual quaternions.

5.1 Algebra

Notation

$$g_{ij} = g_i g_j, \, i \neq j.$$

$$G_0 = g_1 g_2 g_3, G_1 = g_0 g_3 g_2, G_2 = g_0 g_1 g_3, G_3 = g_0 g_2 g_1.$$

$$g_{0123} = g_0 g_1 g_2 g_3$$

Identities

$$g_0^2 = 0, \quad g_1^2 = g_2^2 = g_3^2 = 1$$

$$g_i \cdot g_j = 0 \quad i \neq j$$

$$g_i G_i = g_{0123} = -G_i g_i, \quad i = 0,1,2,3$$

$$g_i G_0 = g_{jk} = G_0 g_i, \quad ijk \text{ is an even permutation of 123}$$

$$g_j G_i = sgn(ijk) g_{0k} = G_i g_j, \quad ijk \text{ a permutation of 123}$$

$$G_j G_0 = g_{0j} = -G_0 G_j \quad j \neq 0$$

$$G_0^2 = -1, \quad G_j^2 = 0 \quad j \neq 0$$

Products $(u, v \ grade \ 1)$

$$uv = u \cdot v + u \wedge v$$

$$u \cdot v = \frac{uv + vu}{2}$$

$$u \wedge v = \frac{uv - vu}{2}$$

$$u \times v = -G_0 (u \wedge v)$$

Wedge Product $(grade(U) = c, \ grade(V) = d)$

$$U \wedge V = \frac{UV + (-1)^{cd} VU}{2}$$

$$s(U \wedge V)s^{-1} = (sUs^{-1}) \wedge (sVs^{-1})$$

5.2 Geometry

Representations

	Clifford Algebra (Plane Model)	Coordinate Representation
Points	$P = G_0 + p_1 G_1 + p_2 G_2 + p_3 G_3$	$P = (p_1, p_2, p_3)$
Vectors	$v = v_1 G_1 + v_2 G_2 + v_3 G_3$	$v = (v_1, v_2, v_3)$
Planes	$\pi = a g_1 + b g_2 + c g_3 + d g_0$	$ax + by + cz + d = 0$
Lines	$l = \pi_2 \wedge \pi_1$	$a_j x + b_j y + c_j z + d_j = 0, \quad j = 1, 2$
	$l = \left(p_2^\perp \wedge p_1^\perp \right)^\perp$	$l = (v, (P - O) \times v)$
	$l = l_{01} g_{01} + l_{02} g_{02} + l_{03} g_{03} + l_{32} g_{32} + l_{13} g_{13} + l_{21} g_{21}$	$l = v + u \varepsilon \quad v \cdot u = 0$
	$0 = l_{01} l_{32} + l_{02} l_{13} + l_{03}\, l_{21}$	

Duality

$$1^\perp = g_{0123} \quad \text{and} \quad g_{0123}^\perp = 1$$

$$g_j^\perp = G_j \quad \text{and} \quad G_j^\perp = -g_j$$

$$g_{ij}^\perp = -g_{0k} \quad \text{and} \quad g_{0k}^\perp = -g_{ij} \ (ijk \ \text{is an even permutation of } 321)$$

$$rr^\perp = g_{0123} \quad \text{for basis multi-vectors}$$

$$\left(r^\perp \right)^\perp = r \quad \text{for even grades} \left(\text{dual quaternions} \right)$$

$$\left(r^\perp \right)^\perp = -r \quad \text{for odd grades}$$

Joins

$$p_1 \vee p_2 = \left(p_2^\perp \wedge p_1^\perp \right)^\perp$$

bivector representing the line joining two distinct points

$$\pi = \left(p_1^\perp \wedge p_2^\perp \wedge p_3^\perp \right)^\perp$$

vector representing the plane containing three non-collinear points

Meets (Intersection Formulas)

$p = \pi_1 \wedge \pi_2 \wedge \pi_3$ *intersection point p of three planes* π_1, π_2, π_3

$p = \pi \wedge l$ *intersection point p between a line l and a plane* π

$l = \pi_2 \wedge \pi_1$ *intersection line l between two planes* π_1, π_2

Incidence Relations

$\pi \wedge p = 0$ *point p lies on plane* π

$l^\perp \wedge p^\perp = 0$ *point p lies on line l*

$\pi \wedge l = 0$ *line l lies on plane* π

$l_1 \wedge l_2 \neq 0$ *lines* l_1, l_2 *are skew*

Distance Formulas

$dist(p, \pi) = \left| (\pi \wedge p)^\perp \right|$ *from a point p to a plane* π *with a unit normal vector*

$$dist(p, \pi) = \left| (p - q)^\perp \cdot v \right|$$

from a point p to a plane π *with a point q and unit normal v*

$$dist^2(p, l) = (p - q)^\perp \cdot (p - q)^\perp - \frac{\left((p - q)^\perp \cdot v^\perp \right)^2}{v^\perp \cdot v^\perp}$$

from a point p to a line l containing a point q and parallel to the direction vector v

$$dist(p, l) = \frac{\left| (p - q)^\perp \times v^\perp \right|}{\left| v^\perp \right|}$$

from a point p to a line l containing a point q and parallel to the direction vector v

5.3 Rotors and Versors

Translation by the distance a in the direction of the unit vector $t = G_0 t$

$$T = 1 - \frac{1}{2}at = e^{-\frac{1}{2}at} \qquad \text{\textit{translation rotor}}$$

$$t = t_1 g_{01} + t_2 g_{02} + t_3 g_{03} \qquad -G_0 t \text{ \textit{is a unit vector}}$$

$$x \mapsto TxT^{-1}$$

Rotation by the angle θ about the line v through the origin G_0

$$R = \cos(\theta/2) + v \sin(\theta/2) = e^{\frac{1}{2}v\theta} \qquad \text{\textit{rotation rotor}}$$

$$v = v_1 g_{32} + v_2 g_{13} + v_3 g_{21} \qquad \text{\textit{unit quaternion}}$$

$$x \mapsto RxR^{-1}$$

Reflection in a plane π with unit normal vector

$$x \mapsto \pi x \pi \qquad \text{\textit{reflection versor}}$$

5.4 Perspective and Pseudo-Perspective

Perspective from the eye E located at the origin G_0 to the plane π at a distance d from the eye perpendicular to the unit normal vector $t^\perp$ dual to the unit vector t.

$$T = 1 + \frac{1}{2d}t = \text{translation rotor with } t = -g_0 t^\perp$$

$$P \mapsto \left(T(P - E)^\perp T^{-1} \right)^\perp$$

Pseudo-perspective with the eye located at $E = G_0 - dt$, a distance d from the origin G_0 in the direction of the unit vector t.

$$T = 1 + \frac{1}{2d}t = \text{translation rotor with } t = -g_0 t^\perp$$

$$P \mapsto \left(TP^\perp T^{-1} \right)^\perp$$

6 Comparisons between Dual Quaternions and the Plane Model of Clifford Algebra

We close by comparing and contrasting some properties of dual quaternions and the plane model of Clifford algebra and by illustrating as well some of the advantages and disadvantages of each approach.

Dimension
- **Dual Quaternions**: 8-Dimensional vector space
- **Clifford Algebra**: 16-Dimensional vector space

Algebra
- $g_{32} \leftrightarrow i$, $g_{13} \leftrightarrow j$, $g_{21} \leftrightarrow k$, $g_{0123} \leftrightarrow \varepsilon$

Geometry

	Plane Model	Dual Quaternions
Points	$p = G_0 + p_1 G_1 + p_2 G_2 + p_3 G_3$	$p = O + (p_1 i + p_2 j + p_3 k)\varepsilon$
Vectors	$v = v_1 G_1 + v_2 G_2 + v_3 G_3$	$v = (p_1 i + p_2 j + p_3 k)\varepsilon$
Planes	$\pi = ag_1 + bg_2 + cg_3 + dg_0$	$\pi = dO\varepsilon + ai + bj + ck$
Lines	$l = l_{32} g_{32} + l_{13} g_{13} + l_{21} g_{21}$	$l = v + u\varepsilon \quad v \cdot u = 0$
	$\quad + l_{01} g_{01} + l_{02} g_{02} + l_{03} g_{03}$	$u, v = \text{quaternions}$
	$l_{01} l_{32} + l_{02} l_{13} + l_{03} l_{21} = 0$	
Meets and Joins		*Dual Plucker Coordinates and*
$l = \pi_2 \wedge \pi_1$		*Plucker Coordinates*
$l = (p_2^{\perp} \wedge p_1^{\perp})^{\perp}$		$(v_1 \times v_2, d_1 \pi_2 - d_2 \pi_1)$
		$\left(v, (p - O) \times v\right)$

Duality

Plane Model	Dual Quaternions
$1^{\perp} = g_{0123}$ and $g_{0123}^{\perp} = 1$	$O^{\#} = O\varepsilon$ and $(O\varepsilon)^{\#} = O$
$g_j^{\perp} = G_j$ and $G_j^{\perp} = -g_j$	$i^{\#} = i\varepsilon$ and $(i\varepsilon)^{\#} = i$
$g_{ij}^{\perp} = -g_{0k}$ and $g_{0k}^{\perp} = -g_{ij}$	$j^{\#} = j\varepsilon$ and $(j\varepsilon)^{\#} = j$
(*ijk* is an even permutation of 321)	$k^{\#} = k\varepsilon$ and $(k\varepsilon)^{\#} = k$
$rr^{\perp} = g_{0123}$ for basis multivectors	$r \cdot r^{\#} = \varepsilon$ for basis elements.
$(r^{\perp})^{\perp} = r$ for even grades	$(r^{\#})^{\#} = r$
$(r^{\perp})^{\perp} = -r$ for odd grades	

Transformations

	Plane Model	Dual Quaternions
Translation	$x \mapsto TxT^{-1}$	$x \mapsto TxT^{\dagger}$
Rotation	$x \mapsto RxR^{-1}$	$x \mapsto RxR^{\dagger}$
Reflection	$x \mapsto \pi x \pi$	$x \mapsto -\pi x^{*} \pi, \quad l \mapsto -\pi l^{\dagger} \pi^{-1}$

Intersections

	Plane Model	Dual Quaternions
Three Planes	$p = \pi_1 \wedge \pi_2 \wedge \pi_3$	$p = O + \dfrac{(v_1 \cdot C_1)(v_2 \times v_3) + (v_2 \cdot C_2)(v_3 \times v_1) + (v_3 \cdot C_3)(v_1 \times v_2)}{\det(v_1 \ v_2 \ v_3)}$
		$\pi_k = v_k - C_k \cdot v_k, \quad k = 1,2,3$
Two Planes	$l = \pi_2 \wedge \pi_1$	$l = (\pi_2 - d_2 O\varepsilon) \times (\pi_1 - d_1 O\varepsilon) + (d_2\pi_1 - d_1\pi_2)\varepsilon$
		$\pi_k = d_k O\varepsilon + v_k, \quad k = 1,2$
Line Plane	$p = \pi \wedge l$	$p = O + \dfrac{v \times u}{|v|^2}\varepsilon + \lambda v\varepsilon, \ \text{where } \lambda = -\dfrac{d\,|v|^2 + \det(v \ u \ v_\pi)}{(v \cdot v_\pi)\,|v|^2}$
		$\pi = dO\varepsilon + v_\pi \quad l = v + u\varepsilon, \ v \cdot u = 0$

Incidence Relations

	Plane Model	Dual Quaternions
point p on plane π	$\pi \wedge p = 0$	$p \cdot \pi = 0$
point p on line l	$l^{\perp} \wedge p^{\perp} = 0$	$(p - O) \times v = u\varepsilon, \quad l = v + u\varepsilon$
line l on plane π	$\pi \wedge l = 0$	$v \cdot v_\pi = 0 \quad \text{and} \quad d\,|v|^2 = \det(u \ v \ v_\pi)$
		$\left(l = v + u\varepsilon \quad \text{and} \quad \pi = dO\varepsilon + v_\pi\right)$
lines l_1, l_2 are skew	$l_1 \wedge l_2 \neq 0$	$\det(v_1 \ v_2 \ P_2 - P_1) \neq 0$
		$\left(l_j = v_j + (P_j - O) \times v_j \quad j = 1,2\right)$

Distance Formulas

	Plane Model	Dual Quaternions
point p to plane π *π has unit normal*	$dist(p,\pi) = \left\lvert (\pi \wedge p)^{\perp} \right\rvert$	$(p \cdot \pi)^{\#}$
point p to line l	$dist(p,l) = \dfrac{\left\lvert (p-q)^{\perp} \times v^{\perp} \right\rvert}{\lvert v^{\perp} \rvert}$ *l* contains the point $q = G_0 + \dfrac{\tilde{u} \times v}{\lvert v \rvert^2}$ and is parallel to the direction vector $v = g_0 l$	$\left\lvert \left(\dfrac{(p-O) \times v - u\varepsilon}{\lvert v \rvert} \right)^{\#} \right\rvert$, $\quad l = v + u\varepsilon$

Points and Translations

- *Dual Quaternions*: Points and translations have the same representation
- *Clifford Algebra*: Points have odd grade; translations have even grade

Computations

- *Dual Quaternions*
  - Simple multiplication rules involving only the quaternions O, i, j, k and one additional new symbol ε.
  - Multiplication table has $5^2 = 25$ entries.

- *Clifford Algebra*
  - Lots of new multiplication formulas involving 16 basis vectors $1, g_j, g_{ij}, G_j, g_{0123}$.
  - Multiplication table has $16^2 = 256$ entries.

Advantages

- *Dual Quaternions*
  - Lower dimension
  - Computations are often simpler and faster with dual quaternions

- *Clifford Algebra*
  - More clearly distinguishes between operators and operands
  - Reflections are much simpler with Clifford algebra
  - Intersections are much simpler with Clifford algebra
  - Incidence relations are sometimes simpler with Clifford algebra

3

The Point Model of Clifford
Algebra for Dual Quaternions

A point-based Clifford algebra model for dual quaternions is proposed by
G. Mullineux and his collaborators [16,21,22] as an alternative to the plane-
based model. We shall call this model *the point model* for short.

1 Algebra

We start in the point model with an orthogonal basis f_0, f_1, f_2, f_3 for R^4
along with an infinitesimal scalar α, so that

$$f_i \cdot f_j = 0 \quad i \neq j,$$

and we set

$$f_0^2 = 1/\alpha, \quad f_1^2 = f_2^2 = f_3^2 = 1. \tag{1.1}$$

Setting $f_0^2 = 1/\alpha$ is a device for avoiding infinities in computations. By
convention, *since α is an infinitesimal, we can set terms multiplied by α to zero
after a computation is complete*. This device is equivalent to taking the
limit as $\alpha \to 0$ after a computation is complete—a standard device in classical calculus.

The main difference between the algebra of the plane model and the
algebra of point model is that in the plane model g_0^2 is set to zero whereas
in the point model f_0^2 is set to $1/\alpha$ which corresponds to infinity. We shall
see in Section 2 that these reciprocals in the algebra have a profound effect
on the geometry, flipping the representations of points and planes, as well
as meets and joins. Transformations, however, especially translation and
rotation rotors are the same in both models.

Multiplication in the point model is associative and distributes through
addition, but as usual multiplication is not commutative since

$$f_j f_i = -f_i f_j, \quad i \neq j, \quad i, j = 0, 1, 2, 3.$$

DOI: 10.1201/9781003398141-16

As in the plane model, two important subalgebras reside inside the point model of Clifford algebra: the quaternions and the dual quaternions. Since $f_1^2 = f_2^2 = f_3^2 = 1$, the entire 8-dimensional Clifford algebra model for quaternions generated by $\{1, f_1, f_2, f_3\}$ is a subalgebra of the point model of Clifford algebra (see Chapter 1, Section 2, Example 2.4.). The quaternions reside inside this model as the even subalgebra generated by $\{1, f_3 f_2, f_1 f_3, f_2 f_1\}$. By definition $f_0^2 = 1/\alpha$, so

$$\left(f_0 f_1 f_2 f_3 \alpha\right)^2 = \alpha \tag{1.2}$$

is an infinitesimal, which can be set to zero after computations are complete. In analogy with the plane model, we shall identify $f_0 f_1 f_2 f_3 \alpha$ with the dual quaternion ε, since $\varepsilon^2 = 0$.

Now the dual quaternions $\tilde{q} = \tilde{q}_1 + \tilde{q}_2 \varepsilon$ are isomorphic to the subalgebra of the point model given by expressions of the form $q = q_1 + q_2 f_{0123} \alpha$, where $q_1, q_2 \in span\left(1, f_3 f_2, f_1 f_3, f_2 f_1\right)$ and $f_{0123} = f_0 f_1 f_2 f_3$.

To simplify our presentation, we will adopt the following notation similar to the notation in plane model but with f replacing g and F replacing G:

$$f_{ij} = f_i f_j, \quad i \neq j.$$

$$F_0 = f_1 f_2 f_3, \quad F_1 = f_0 f_3 f_2, \quad F_2 = f_0 f_1 f_3, \quad F_3 = f_0 f_2 f_1.$$

$$f_{0123} = f_0 f_1 f_2 f_3,$$

Also as in the plane model, we shall make use of the following five identities (see Exercise 2):

$$f_i F_i = f_{0123} = -F_i f_i, \quad i = 0, 1, 2, 3. \tag{1.3}$$

$$f_i F_0 = f_{jk} = F_0 f_i, \quad ijk \text{ an even permutation of 123} \tag{1.4}$$

$$f_j F_i = sgn(ijk) f_{0k} = F_i f_j, \quad ijk \text{ a permutation of 123} \tag{1.5}$$

$$F_i F_j = -F_j F_i \qquad j \neq i \tag{1.6}$$

$$F_j^2 \alpha = F_0^2 = -1. \quad j \neq 0 \tag{1.7}$$

With this notation we have the following dictionary for translating between the algebra of dual quaternions and the algebra of the point model.

TABLE 3.1

Translating between the Basis Vectors in the Algebra of Dual Quaternions and the Blades in the Subalgebra of the Point Model Isomorphic to the Dual Quaternions

Dual Quaternions	Point Model
O	1
i	f_{32}
j	f_{13}
k	f_{21}
ε	$f_{0123}\alpha$

The symbol O in the space of dual quaternions plays two roles: algebraic and geometric. Algebraically O represents the identity for multiplication; geometrically O represents the origin. As in the plane model, these two roles are split apart in the point model. In the point model, 1 is the identity for multiplication, whereas, as we shall see shortly, f_0 represents the origin. Splitting apart these two roles of O facilitates the separation of operators and operands—one of the advantages of Clifford algebra over the dual quaternions.

Exercises

1. Show that the dual real numbers reside inside the point model of Clifford algebra as the subalgebra generated by $\{1, f_{0123}\alpha\}$.

2. Verify identities (1.3)–(1.7).

3. In this exercise we investigate the inverses of the basis multivectors in the point model. Show that:

   a. $f_0^{-1} = f_0\alpha \quad f_j^{-1} = f_j \quad\quad j \neq 0$

   b. $F_0^{-1} = -F_0 \quad F_j^{-1} = -F_j\alpha \quad\quad j \neq 0$

   c. $f_{0k}^{-1} = -f_{0k}\alpha \quad f_{ij}^{-1} = -f_{ij} \quad i, j \neq 0$

4. Show that

   a. $F_0 f_0 F_0^{-1} = -f_0$

   b. $F_0 f_j F_0^{-1} = f_j \quad\quad j = 1, 2, 3$

5. Using the results of Exercise 4, show that:

   a. $F_0 f_{0j} F_0^{-1} = -f_{0j}$ $j = 1, 2, 3$

   b. $F_0 f_{ij} F_0^{-1} = f_{ij}$ $i, j = 1, 2, 3$

   c. $F_0 F_0 F_0^{-1} = F_0$

   d. $F_0 F_j F_0^{-1} = -F_j$ $j = 1, 2, 3$

   e. $F_0 f_{0123} F_0^{-1} = -f_{0123}$

6. Compare each of the preceding exercises to the corresponding exercises in Chapter 2, Section 1.

7. Show that:

   a. $f_0 f_0 f_0^{-1} = f_0$

   b. $f_0 f_j f_0^{-1} = -f_j$ $j = 1, 2, 3$

8. Using the results of Exercise 7, show that:

   a. $f_0 f_{0k} f_0^{-1} = -f_{0k}$ $k \neq 0$

   b. $f_0 f_{ij} f_0^{-1} = f_{ij}$ $i, j \neq 0$

   c. $f_0 F_0 f_0^{-1} = -F_0$

   d. $f_0 F_j f_0^{-1} = F_j$ $j \neq 0$

   e. $f_0 f_{0123} f_0^{-1} = -f_{0123}$

2 Geometry

The representation of geometry in the point model is quite natural. Points and vectors are the basic elements: lines are represented as the join of two points or the join of a point and a vector, and planes are represented as the join of three points, three vectors, or a point and two vectors.

Geometry is represented in terms of the basis multivectors: f_0 represents the origin, and the other three basis vectors f_1, f_2, f_3 represent unit vectors along the x, y, z-coordinate axes. For arbitrary points and vectors, the point model adopts the standard representation of homogeneous coordinates using linear combinations of f_0, f_1, f_2, f_3 to represent points and vectors. Lines are represented by bivectors. The bivectors f_{32}, f_{13}, f_{21} represent lines along the x, y, z -coordinate axes. Lines at infinity are represented by the other three basis bivectors. The bivector f_{01} represents the ideal line in the $x = 0$ plane; similarly, f_{02} represents the ideal line in the $y = 0$ plane, and f_{03} represents the ideal line in the $z = 0$ plane. Finally, trivectors represent planes. The trivector F_0 represents the plane at infinity, while F_1 represents the yz-plane $(x = 0)$, F_2 represents the xz-plane $(y = 0)$, and F_3 represents the xy-plane $(z = 0)$.

These definitions describe the geometry associated with the basis multivectors of the point model of the Clifford algebra for the dual quaternions. Although these definitions seem quite natural, we need to understand how these definitions relate back to the geometry associated with the dual quaternions described in Part I. Next we will explore these connections and extend these definitions to representations for arbitrary vectors, points, lines, and planes in 3-dimensions.

2.1 Points and Vectors

To see how our representation for points and vectors in this point model of Clifford algebra relates to the representation of points and vectors in the space of dual quaternions, recall first the representation of points $\tilde{p}$ with coordinates (p_1, p_2, p_3) in the space of dual quaternions:

$$\tilde{p} = O + (p_1 i + p_2 j + p_3 k)\varepsilon$$

The corresponding dual quaternion in the point model of Clifford algebra is (see Table 3.1)

$$\tilde{p} = 1 + (p_1 f_{32} + p_2 f_{13} + p_3 f_{21}) f_{0123}\alpha.$$

Notice, however, that $\tilde{p}$ is an element of mixed grade and contains the symbol α. To generate an element of fixed grade and to eliminate α, we can multiply $\tilde{p}$ on the left by f_0 :

$$f_0 \tilde{p} = f_0 + (p_1 f_{32} + p_2 f_{13} + p_3 f_{21}) f_{123} = f_0 + p_1 f_1 + p_2 f_2 + p_3 f_3.$$

The expression

$$p = f_0 + p_1 f_1 + p_2 f_2 + p_3 f_3 = f_0 \tilde{p} \tag{2.1}$$

is an element of grade 1, so in the point model it is natural to use these grade 1 elements to represent the points with coordinates (p_1, p_2, p_3). Thus f_0 represents the origin and the other three basis vectors f_1, f_2, f_3

represent unit vectors along the x, y, z-coordinate axes. In particular, any vector $v = (v_1, v_2, v_3)$ can be expressed as

$$v = v_1 f_1 + v_2 f_2 + v_3 f_3 = f_0 \tilde{v} \tag{2.2}$$

where

$$\tilde{v} = (v_1 f_{32} + v_2 f_{13} + v_3 f_{21}) f_{0123} \alpha \leftrightarrow (v_1 i + v_2 j + v_3 k)\varepsilon$$

represents the vector with coordinates (v_1, v_2, v_3) in the space of dual quaternions. Notice how multiplying by f_0 mediates between the dual quaternion and this Clifford algebra representations for points and vectors.

2.2 Planes

In the space of dual quaternions, the planes $\tilde{\pi}$ corresponding to the equations

$$ax + by + cz + d = 0$$

are represented by expressions of the form

$$\tilde{\pi} = ai + bj + ck + dO\varepsilon.$$

The corresponding dual quaternions in the point model of Clifford algebra are (see Table 3.1)

$$\tilde{\pi} = af_{32} + bf_{13} + cf_{21} + df_{0123}\alpha.$$

Notice again that $\tilde{\pi}$ is an element of mixed grade and contains the symbol α. To generate an element of fixed grade and to eliminate α, we can once again multiply on the left by f_0:

$$f_0 \tilde{\pi} = aF_1 + bF_2 + cF_3 + dF_0.$$

The expression

$$\pi = aF_1 + bF_2 + cF_3 + dF_0 = f_0 \tilde{\pi} \tag{2.3}$$

is an element of grade 3, so it is natural to use these grade 3 elements to represent the planes corresponding to the equation

$$ax + by + cz + d = 0.$$

Thus F_0 represents the plane at infinity and the other three blades F_1, F_2, F_3 represent the yz, xz, xy-coordinate planes. In particular, when $d = 0$ the plane π passes through the origin. Notice that in the point model, the coordinate planes are represented by the joins of one point and two vectors, whereas the plane at infinity is represented as the join of three vectors.

Exercises

1. Show that the join of three vectors represents the plane at infinity.
2. Show that the join of the origin and two vectors represents a plane through the origin.

2.2.1 Incidence Relation: Point on Plane

Next, we shall show how the formula for the incidence relation between a point and a plane in this point model of Clifford algebra can be derived directly from the formula for this incidence relation in the algebra of dual quaternions. Recall from Part I, Proposition 3.1 that in the space of dual quaternions, the point $\tilde{p}$ lies on the plane $\tilde{\pi}$ if and only if $\tilde{p} \cdot \tilde{\pi} = 0$.

Lemma 2.1

Suppose that $p = f_0 \tilde{p}$ and $\pi = f_0 \tilde{\pi}$. Then

$$\text{a.} \quad \tilde{p} = f_0 \alpha p \quad \text{and} \quad \tilde{\pi} = f_0 \alpha \pi \tag{2.4}$$

$$\text{b.} \quad \tilde{p}^* = p f_0 \alpha \quad \text{and} \quad \tilde{\pi}^* = -\pi f_0 \alpha \tag{2.5}$$

where star denotes the quaternion conjugate.

Proof: Part a follows immediately because $f_0^2 = 1/\alpha$. To prove part b, observe that

$$p f_0 \alpha = (f_0 + p_1 f_1 + p_2 f_2 + p_3 f_3) f_0 \alpha = 1 - (p_1 f_{32} + p_2 f_{13} + p_3 f_{21}) f_{0123} \alpha = \tilde{p}^*$$

$$-\pi f_0 \alpha = (a F_1 + b F_2 + c F_3 + d F_0)(-f_0 \alpha) = -(a f_{32} + b f_{13} + c f_{21}) + d f_{0123} \alpha = \tilde{\pi}^*. \ \Diamond$$

Proposition 2.2

Let p be a point and π be a plane in the point model of Clifford algebra, and let $\tilde{p}$ and $\tilde{\pi}$ be the corresponding point and plane in the space of dual quaternions. Then

$$\tilde{p} \cdot \tilde{\pi} = (p \wedge \pi)\alpha. \tag{2.6}$$

Proof: By Lemma 2.1 and the definition of the dot product in the space of dual quaternions (Part I, Equation 2.1)

$$\tilde{p} \cdot \tilde{\pi} = \frac{\tilde{p}\tilde{\pi}^* + \tilde{\pi}\tilde{p}^*}{2} = \frac{\left(f_0\alpha p\right)\left(-\pi f_0\alpha\right) + \left(f_0\alpha\pi\right)\left(pf_0\alpha\right)}{2} = f_0\alpha\left(\frac{\pi p - p\pi}{2}\right)f_0\alpha.$$

Therefore by Chapter 1, Proposition 3.3

$$\tilde{p} \cdot \tilde{\pi} = f_0\alpha\left(\pi \wedge p\right)f_0\alpha.$$

Now $\mathrm{grade}(p) = 1$ and $\mathrm{grade}(\pi) = 3$, so $\mathrm{grade}(\pi \wedge p) = 4$. Hence $\pi \wedge p$ is a multiple of the pseudoscalar f_{0123}. Thus there is a constant c such that $\pi \wedge p = cf_{0123}$. Therefore

$$\tilde{p} \cdot \tilde{\pi} = f_0\alpha\left(\pi \wedge p\right)f_0\alpha = f_0\alpha\left(cf_{0123}\right)f_0\alpha = \left(cF_0\right)f_0\alpha = -cf_{0123}\alpha = \left(p \wedge \pi\right)\alpha. \lozenge$$

Remark 2.3

Although in Proposition 2.2 $p \wedge \pi$ is multiplied by the infinitesimal α, the expression $\left(p \wedge \pi\right)\alpha$ is not zero because $\left(p \wedge \pi\right)\alpha = -cf_{0123}\alpha$ and $f_{0123}\alpha$ is identified with the dual quaternion $\varepsilon \neq 0$. See too Exercise 1.

Theorem 2.4

A point p lies on the plane π if and only if $p \wedge \pi = 0$.

Proof: Let $\tilde{p}$ and $\tilde{\pi}$ be the point and plane in the space of dual quaternions corresponding to the point p and plane π in the point model of Clifford algebra. Then the point p lies on the plane π if and only if the point $\tilde{p}$ lies on the plane $\tilde{\pi}$. Moreover the point $\tilde{p}$ lies on the plane

$\tilde{\pi}$ if and only if $\tilde{p} \cdot \tilde{\pi} = 0$. But by Proposition 2.2 $\tilde{p} \cdot \tilde{\pi} = 0$ if and only if $(p \wedge \pi)\alpha = 0$. However α is invertible, since $f_0^2 = 1/\alpha$. Hence the point p lies on the plane π if and only if $p \wedge \pi = 0$. ◊

Corollary 2.5

Let p_1, p_2, p_3 be three noncollinear points. Then the plane $\pi = p_1 \wedge p_2 \wedge p_3$ contains all three points p_1, p_2, p_3.

Proof: This result follows immediately from Theorem 2.4 because

$$p_j \wedge \pi = p_j \wedge (p_1 \wedge p_2 \wedge p_3) = 0 \quad \text{for } j = 1, 2, 3.$$

Therefore the plane $\pi = p_1 \wedge p_2 \wedge p_3$ contains all three points p_1, p_2, p_3. ◊

Exercise

1. Let $p = f_0 + xf_1 + yf_2 + zf_3$ be a point and let $\pi = aF_1 + bF_2 + cF_3 + dF_0$ be a plane in the point model of Clifford algebra.

   a. Using Equations (1.3)–(1.5), show that

   $$\frac{\pi p - p\pi}{2} = -(ax + by + cz + d) f_{0123}.$$

   b. Using properties of the wedge product, show that

   $$\pi \wedge p = -(ax + by + cz + d) f_{0123}.$$

   c. Using part a or part b, show that

   $$(\pi \wedge p)\alpha = -(ax + by + cz + d) f_{0123}\alpha.$$

   d. Conclude from part *a* or part *b* that the point p lies on the plane π if and only if $p \wedge \pi = 0$.

2.3 Duality

We have already encountered duality twice in Clifford algebra: once in the Clifford algebra for the quaternions, and once in the plane model of Clifford algebra for the dual quaternions. Here we shall develop duality formulas for the point model, identical in form to the formulas for duality for the plane model. We shall then use these duality formulas for three specific purposes in the point model for the dual quaternions:

i. to construct lines as both the join of two points (Section 2.4.1) and the meet of two planes (Section 2.4.4);

ii. to apply translation, rotation, and reflection to lines (Sections 3.1, 3.2, 3.3);

iii. to compute perspective and pseudo-perspective (Section 3.4).

2.3.1 Duality in the Point Model

The algebra of the point model differs from the algebra of the plane model because $f_0^2 = 1/\alpha$ whereas $g_0^2 = 0$. But we observed in Chapter 1, Section 4 that the formulas for duality are independent of the formulas for the squares of the basis vectors. Therefore, the formulas for duality in the point model are identical in form to the formulas for duality in the plane model; we need only change the g's to f's. Thus, we can immediately write down the duality formulas for the basis multivectors simply by copying these formulas from the plane model and replacing g's by f's and G's by F's:

$$1^\perp = f_{0123} \quad \text{and} \quad f_{0123}^\perp = 1 \tag{2.7}$$

$$f_j^\perp = F_j \quad \text{and} \quad F_j^\perp = -f_j \tag{2.8}$$

$$f_{ij}^\perp = -f_{0k} \quad \text{and} \quad f_{0k}^\perp = -f_{ij} \quad (ijk \text{ is an even permutation of 321}) \tag{2.9}$$

For the basis multivectors:

$$rr^\perp = f_{0123} \quad \text{for all grades} \tag{2.10}$$

$$(r^\perp)^\perp = r \quad \text{for even grades } (\text{dual quaternions}) \tag{2.11}$$

$$(r^\perp)^\perp = -r \quad \text{for odd grades} \tag{2.12}$$

We then extend to arbitrary multivectors by linearity.

Geometric properties of duality carry over from the plane model to the point model: vectors and normal vectors are dual, affine lines and lines at infinity are dual, and the origin f_0 and the plane at infinity F_0 are dual.

Exercises

1. Show that in the point model duality is mediated by multiplication with pseudoscalars. In particular, show that:

    a. $1^\perp = 1(f_{0123})$ and $f_{0123}^\perp = f_{0123}(f_{0123}\alpha)$.

    b. $f_j^\perp = f_j(f_{0123})$ and $F_j^\perp = F_j(-f_{0123}\alpha)$ $\quad j = 1,2,3$

    c. $f_0^\perp = f_0(f_{0123}\alpha)$ and $F_0^\perp = F_0(-f_{0123})$

    d. $f_{ij}^\perp = f_{ij}(-f_{0123})$ and $f_{0k}^\perp = f_{0k}(-f_{0123}\alpha)$ $\quad i,j,k \neq 0$.

2. Show that, in general, even for basis multivectors:

    a. $(r \wedge s)^\perp \neq r^\perp \wedge s^\perp$.

    b. $(r \wedge s)^\perp \neq s^\perp \wedge r^\perp$.

    c. $r \wedge s^\perp \neq r^\perp \wedge s$.

3. Let u, v be vectors. Show that:

    a. $(u \cdot v)^\perp = v \wedge u^\perp$.

    b. $(u \wedge v)^\perp = f_0(u \times v)$.

    c. Reconcile the difference between the equation in part b and Equation (4.1) in Chapter 1.

4. Let u, v be vectors. Show that:

$$-\left(\frac{u^\perp v^\perp + v^\perp u^\perp}{2}\right)\alpha = u \cdot v.$$

5. Let π be a plane with a unit normal vector v and containing a point q, and let p be an arbitrary point. Show that:

    a. $dist(p,\pi) = \left|(p-q) \cdot v^\perp\right|$.

    b. $dist(p,\pi) = \left|(\pi \wedge p)^\perp\right|$.

6. Let π be the plane with normal vector v passing through the point p. Show that:

    a. $\pi = v - (p \wedge v)^\perp F_0$.

2.4 Lines

Lines are typically represented in the space of dual quaternions using either Plucker coordinates or dual Plucker coordinates. Since we are working in the point model of Clifford algebra for dual quaternions, it is natural here to use Plucker coordinates to represent lines. Moreover, since points have grade 1 and planes have grade 3, we should expect that, as in the plane model, lines would have grade 2. In addition, since the wedge product of three points is the plane containing the three points (join), it is again natural to expect that the wedge product $p_1 \wedge p_2$ of two points p_1, p_2 would represent the line joining these two points. This formula is almost correct, but not quite. In fact, we shall see that the join of two points is constructed by reversing their order and taking the dual of the wedge product.

2.4.1 Lines as the Join of Two Points

Let p_1 and p_2 be two distinct points. We shall show that the join $p_1 \vee p_2 = (p_2 \wedge p_1)^{\perp}$ generates the Plucker coordinates [Part I, Section 1.3.3.1] of the line l joining the points p_1 and p_2.

To see how, let

$$p_1 = f_0 + a_1 f_1 + a_2 f_2 + a_3 f_3$$

$$p_2 = f_0 + b_1 f_1 + b_2 f_2 + b_3 f_3.$$

Then

$$p_2 \wedge p_1 = (a_1 - b_1) f_{01} + (a_2 - b_2) f_{02} + (a_3 - b_3) f_{03}$$
$$+ (b_3 a_2 - b_2 a_3) f_{32} + (b_1 a_3 - b_3 a_1) f_{13} + (b_2 a_1 - b_1 a_2) f_{21}$$

$$l = (p_2 \wedge p_1)^{\perp} = (b_1 - a_1) f_{32} + (b_2 - a_2) f_{13} + (b_3 - a_3) f_{21}$$
$$+ (a_3 b_2 - a_2 b_3) f_{01} + (a_1 b_3 - a_3 b_1) f_{02} + (a_2 b_1 - a_1 b_2) f_{03}$$

Thus $(p_2 \wedge p_1)^{\perp}$ has six coefficients relative to the six basis bivectors f_{ij} of grade 2. These six coefficients are precisely the six Plucker coordinates of the line joining the two points p_1 and p_2. To see these Plucker coordinates explicitly, split l into the sum two bivectors:

$$\tilde{v} = (b_1 - a_1) f_{32} + (b_2 - a_2) f_{13} + (b_3 - a_3) f_{21}$$

$$\tilde{u} = (a_3 b_2 - a_2 b_3) f_{01} + (a_1 b_3 - a_3 b_1) f_{02} + (a_2 b_1 - a_1 b_2) f_{03}$$

$$l = \tilde{v} + \tilde{u}$$

and set

$$v = F_0 \tilde{v}$$

$$u = -f_0 \alpha \tilde{u}$$

Then

$$v = (b_1 - a_1) f_1 + (b_2 - a_2) f_2 + (b_3 - a_3) f_3 = p_2 - p_1$$

$$u = (a_2 b_3 - a_3 b_2) f_1 + (a_3 b_1 - a_1 b_3) f_2 + (a_1 b_2 - a_2 b_1) f_3 = (p_1 - f_0) \times (p_2 - f_0)$$

$$= (p_1 - f_0) \times (p_2 - p_1) = (p_1 - f_0) \times v$$

Thus the pair

$$(v, u) = \left(p_2 - p_1, (p_1 - f_0) \times (p_2 - p_1) \right)$$

are precisely the classical Plucker coordinates of the line l joining the points p_1 and p_2. Notice that as required: u·v=0. (See Exercise 5c.) We can now express the bivector l directly in terms of these Plucker coordinates, since

$$l = \tilde{v} + \tilde{u} = -(F_0 v + f_0 u) = -\left(F_0 (p_2 - p_1) + f_0 ((p_1 - f_0) \times (p_2 - p_1)) \right). \quad (2.13)$$

Notice too that the direction of the line l is parallel to the vector

$$v = F_0 \tilde{v} = -(f_0 \wedge l)^{\perp}. \quad (2.14)$$

Exercises

1. Show that the join $p_1 \vee p_2 = (p_2 \wedge p_1)^{\perp}$ is an antisymmetric, bilinear function. That is, show that for any three points p, q, r:
   a. $p \vee q = -q \vee p$
   b. $(ap + bq) \vee r = a(p \vee r) + b(q \vee r)$
   c. $p \vee (cq + dr) = c(p \vee q) + d(p \vee r)$

2. Show that the line represented as the join of two points is independent of choice of points on the line. That is, show that if p_1, p_2 and q_1, q_2 lie on the same line, then there is a constant $\lambda \neq 0$ such that $(q_2 \wedge q_1)^\perp = \lambda (p_2 \wedge p_1)^\perp$. (Hint: Express q_1, q_2 in terms of p_1, p_2 and apply Exercise 1.)

3. Let p, q be two distinct points. Show that the bivector $l = ((q - p) \wedge p)^\perp$ represents the line through the point p parallel to the direction $q - p$.

4. Let p be a point and let v be a vector. Show that the bivector $l = (v \wedge p)^\perp$ represents the line through the point p parallel to the vector v.

5. Let

$$p_1 = f_0 + a_1 f_1 + a_2 f_2 + a_3 f_3$$

$$p_2 = f_0 + b_1 f_1 + b_2 f_2 + b_3 f_3$$

$$\tilde{v} = (b_1 - a_1) f_{32} + (b_2 - a_2) f_{13} + (b_3 - a_3) f_{21}$$

$$\tilde{u} = (a_3 b_2 - a_2 b_3) f_{01} + (a_1 b_3 - a_3 b_1) f_{02} + (a_2 b_1 - a_1 b_2) f_{03}$$

$$v = F_0 \, \tilde{v}$$

$$u = -f_0 \alpha \tilde{u}$$

Show that:

a. $\tilde{v}$ is a line through the origin in the direction $p_2 - p_1$.

b. $\tilde{u}$ is a line at infinity.

c. $u \cdot v = 0$

6. Let $l = (p_2 \wedge p_1)^\perp$ be a bivector representing a line as the join of two points, and let $v = -(f_0 \wedge l)^\perp = p_2 - p_1$. Show that:

a. $dist^2(p, l) = (p - p_1) \cdot (p - p_1) - \dfrac{((p - p_1) \cdot v)^2}{v \cdot v}$.

b. $dist(p, l) = \dfrac{|(p - p_1) \times v|}{|v|}$.

2.4.2 Lines as Bivectors

We have seen that bivectors can represent lines as the join of two points.
But not all bivectors represent lines: as with Plucker coordinates there is
a constraint. Recall from Section 2.4.1 that the join of two points is represented by expressions of the form

$$l = \tilde{v} + \tilde{u},$$

where

$$\tilde{v} = l_{32} f_{32} + l_{13} f_{13} + l_{21} f_{21}$$

$$\tilde{u} = l_{01} f_{01} + l_{02} f_{02} + l_{03} f_{03}.$$

But there is a constraint. Let

$$v = F_0 \, \tilde{v} = l_{32} f_1 + l_{13} f_2 + l_{21} f_3$$

$$u = -f_0 \, \alpha \tilde{u} = l_{01} f_1 + l_{02} f_2 + l_{03} f_3.$$

Then by Section 2.4.1, Exercise 5c, $u \cdot v = 0$. Thus we require that

$$l_{01} l_{32} + l_{02} l_{13} + l_{03} l_{21} = u \cdot v = 0.$$

Therefore if a bivector l represents a line (or equivalently the join of two
points), then we must have

$$l_{01} l_{32} + l_{02} l_{13} + l_{03} l_{21} = 0.$$

Hence if $l_{01} l_{32} + l_{02} l_{13} + l_{03} l_{21} \neq 0$, then the bivector l does not represent a line.
But if $l_{01} l_{32} + l_{02} l_{13} + l_{03} l_{21} = 0$, then we can always find two points p_1, p_2
such that

$$l = l_{32} f_{32} + l_{13} f_{13} + l_{21} f_{21} + l_{01} f_{01} + l_{02} f_{02} + l_{03} f_{03} = p_1 \vee p_2.$$

To see how, once again split the bivector l into the sum two bivectors:

$$\tilde{v} = l_{32} f_{32} + l_{13} f_{13} + l_{21} f_{21}$$

$$\tilde{u} = l_{01} f_{01} + l_{02} f_{02} + l_{03} f_{03}$$

$$l = \tilde{v} + \tilde{u}.$$

To find the points p_1, p_2, we shall construct vectors w_1, w_2 such that

$$v = l_{32} f_1 + l_{13} f_2 + l_{21} f_3 = w_2 - w_1$$

$$u = l_{01} f_1 + l_{02} f_2 + l_{03} f_3 = w_2 \times w_1.$$

Then we would have

$$v \cdot u = l_{01} l_{32} + l_{02} l_{13} + l_{03} l_{21} = 0$$

$$\tilde{v} = -(f_0 v)^\perp = \left(f_0 (w_1 - w_2)\right)^\perp$$

$$\tilde{u} = f_0 u = f_0 (w_2 \times w_1) = (w_2 \wedge w_1)^\perp \quad \text{(see Exercise 2)}.$$

Now we could set

$$p_1 = f_0 + w_1$$
$$p_2 = f_0 + w_2$$

and we would have

$$\left(p_2 \wedge p_1\right)^\perp = \left((f_0 + w_2) \wedge (f_0 + w_1)\right)^\perp$$

$$= \left(f_0 (w_1 - w_2)\right)^\perp + (w_2 \wedge w_1)^\perp = \tilde{v} + \tilde{u} = l.$$

The following lemma shows that we can always find such vectors w_1, w_2, whenever $v \cdot u = 0$.

Lemma 2.6

Let v, u be two vectors with $v, u \neq 0$ and $v \cdot u = 0$. Set $w_1 = \dfrac{u \times v}{|v|^2}$ and set $w_2 = w_1 + v$. Then

i. $v = w_2 - w_1$

ii. $u = w_2 \times w_1$

Proof. Part i is immediate by construction. To prove part ii, observe that since $v \cdot u = 0$, the vectors v and u are orthogonal. Therefore, the vectors $u, v, u \times v$ are mutually orthogonal and form a right-handed system. Hence $w_2 \times w_1 = v \times w_1 = v \times \dfrac{u \times v}{v \cdot v}$ is parallel to u. It remains only to show that $|w_2 \times w_1| = \left| v \times \dfrac{u \times v}{v \cdot v} \right| = |u|$. But since v and u are orthogonal,

$$\left|\frac{u \times v}{|v|^2}\right| = \frac{|u| \times |v|}{|v|^2} = \frac{|u|}{|v|}.$$

Therefore since $u \times v$ and v are orthogonal

$$|w_2 \times w_1| = \left|v \times \frac{u \times v}{v \cdot v}\right| = |v|\left|\frac{u \times v}{|v|^2}\right| = |v|\frac{|u|}{|v|} = |u|.$$

Thus we conclude that $w_2 \times w_1 = u$. $\lozenge$

Corollary 2.7

Let $l = l_{32} f_{32} + l_{13} f_{13} + l_{21} f_{21} + l_{01} f_{01} + l_{02} f_{02} + l_{03} f_{03}$ be a bivector represent-ing a line. Set

- $v = l_{32} f_1 + l_{13} f_2 + l_{21} f_3$

- $u = l_{01} f_1 + l_{02} f_2 + l_{03} f_3$.

Then

i. $p = f_0 + \dfrac{u \times v}{|v|^2}$ *is a point on the line l*

ii. $v = -(f_0 \wedge l)^{\perp}$ *is a vector parallel to the line l.*

Thus the direction of the line l depends only on quaternion term $\tilde{v} = l_{32} f_{32} + l_{13} f_{13} + l_{21} f_{21}$. Two lines that share the same quaternion term $\tilde{v}$ are necessarily parallel (see too Exercise 6).

Proof: This result follows from Lemma 2.6 and the discussion pre-ceding Lemma 2.6. $\lozenge$

In the preceding discussion, we have tacitly assumed that $\tilde{u}, \tilde{v} \neq 0$. Let's now consider each of these two special cases in turn.

Case 1: $l = l_{32} f_{32} + l_{13} f_{13} + l_{21} f_{21}. (\tilde{u} = 0)$
In this case, l is a pure quaternion. Let

$$v = l_{32} f_1 + l_{13} f_2 + l_{21} f_3 = lF_0$$

$$p_1 = f_0$$

$$p_2 = f_0 + v.$$

Then

$$\left(p_2 \wedge p_1\right)^{\perp} = \left(\left(f_0 + v\right) \wedge f_0\right)^{\perp} = \left(v \wedge f_0\right)^{\perp} = \left(l_{32} f_{10} + l_{13} f_{20} + l_{21} f_{30}\right)^{\perp}$$

$$= l_{32} f_{32} + l_{13} f_{13} + l_{21} f_{21} = l$$

Therefore, the bivector l represents the line through the origin f_0 parallel to the vector $v = lF_0$.

Case 2: $l = l_{01} f_{01} + l_{02} f_{02} + l_{03} f_{03}$ $(\tilde{v} = 0)$
In this case, let

$$v = l_{01} f_1 + l_2 f_2 + l_3 f_3$$

Choose two vectors w_1, w_2 orthogonal to v so that $v = w_2 \times w_1$. Then

$$\left(w_2 \wedge w_1\right)^{\perp} = f_0 (w_2 \times w_1) = f_0 v = l$$

Therefore, since l is the join of two vectors rather than the join of two points, the bivector l does not represent an affine line; rather l represents a line at infinity in the plane containing the two vectors w_1, w_2 orthogonal to the vector $w_2 \times w_1$.

Exercises

1. Consider a bivector

$$l = l_{32} f_{32} + l_{13} f_{13} + l_{21} f_{21} + l_{01} f_{01} + l_{02} f_{02} + l_{03} f_{03}.$$

   Let

$$\tilde{v} = l_{01}(f_0 \wedge f_1) + l_{02}(f_0 \wedge f_2) + l_{03}(f_0 \wedge f_3)$$

$$\tilde{u} = l_{32}(f_3 \wedge f_2) + l_{13}(f_1 \wedge f_3) + l_{21}(f_2 \wedge f_1)$$

   Show that the bivector l represents a line if and only if $\tilde{v} \wedge \tilde{u} = 0$.

2. Let w_2, w_1 be vectors. Show that

   a. $(w_2 \wedge w_1)^{\perp} = f_0(w_2 \times w_1)$.

   b. $w_2 \wedge w_1$ represents a line through the origin parallel to the vector $w_1 \times w_2$.

3. Consider a line represented by the bivector

$$l = l_{32} f_{32} + l_{13} f_{13} + l_{21} f_{21} + l_{01} f_{01} + l_{02} f_{02} + l_{03} f_{03}.$$

Show that

$$v = l_{32} f_1 + l_{13} f_2 + l_{21} f_3 = -(f_0 \wedge l)^{\perp}$$

$$u = l_{01} f_1 + l_{02} f_2 + l_{03} f_3 = \left(f_0 \wedge l^{\perp}\right)^{\perp}$$

4. Show that:
   a. f_{32} represents the x-axis
   b. f_{13} represents the y-axis
   c. f_{21} represents the z-axis.

5. Show that:
   a. f_{01} represents the ideal line in the $x = 0$ plane
   b. f_{02} represents the ideal line in the $y = 0$ plane
   c. f_{03} represents the ideal line in the $z = 0$ plane.

6. Consider two lines l, l' represented by the bivectors

$$l = l_{32} f_{32} + l_{13} f_{13} + l_{21} f_{21} + l_{01} f_{01} + l_{02} f_{02} + l_{03} f_{03}$$

$$l' = l'_{32} f_{32} + l'_{13} f_{13} + l'_{21} f_{21} + l'_{01} f_{01} + l'_{02} f_{02} + l'_{03} f_{03}$$

Show that the lines l, l' are parallel if and only if there is a constant $\lambda \neq 0$ such that

$$l'_{32} f_{32} + l'_{13} f_{13} + l'_{21} f_{21} = \lambda \left(l_{32} f_{32} + l_{13} f_{13} + l_{21} f_{21} \right).$$

2.4.3 Incidence Relations for Lines

Given a bivector l representing a line, five problems naturally arise:

1. Find two points (not necessarily unique) whose join is the line l.
2. Determine whether a point p lies on the line l.
3. Determine whether the line l lies on the plane π.
4. Find the intersection point p of the line l with a plane π.
5. Determine whether two lines l_1, l_2 in 3-dimensions intersect.

We have already seen in the preceding section how to solve the first problem by using Corollary 2.7. We will see shortly that we can use the solution to the first problem to solve the second problem. We shall defer the solutions to the third, fourth, and fifth problems to the next section (see Exercises 6 and 7 in Section 2.4.4).

Theorem 2.8

Let l be an affine line and let p be a point.

1. *The point p lies on the line l if and only if $p \wedge l^{\perp} = 0$.*

2. *If the point p does not lie on the line l, then $p \wedge l^{\perp}$ is a plane containing both the line l and the point p.*

Proof: Since the bivector l represents an affine line, there are two points p_1, p_2 such that $l = (p_2 \wedge p_1)^{\perp}$.

1. The point p lies on the affine line $l = (p_2 \wedge p_1)^{\perp}$ if and only if there is a constant λ such that $p = (1 - \lambda)p_1 + \lambda p_2$. Hence p lies on the line l if and only if

$$p \wedge l^{\perp} = \big((1 - \lambda)p_1 + \lambda p_2\big) \wedge (p_2 \wedge p_1) = 0.$$

2. If the point p does not lie on the line l, then $p \wedge l^{\perp} = p \wedge p_2 \wedge p_1 \neq 0$. Therefore by Corollary 2.5, the plane $\pi = p \wedge l^{\perp} = p \wedge p_2 \wedge p_1$ contain all three points p, p_1, p_2, so the plane $\pi = p \wedge l^{\perp}$ is a plane containing both the line l and the point p. $\lozenge$

Exercises

1. Let π be a plane and let l be a line. Explain why $\pi \wedge l = 0$ does not prove that the line l lies on the plane π.

2. The bivector f_{21} represents the z-axis, and the expression $f_0 + \lambda f_3$ represents points along the z-axis. Confirm part 1 of Theorem 2.8 by verifying that $(f_0 + \lambda f_3) \wedge f_{21}^{\perp} = 0$.

2.4.4 Lines as the Intersection of Two Planes

The join of two points determines a line. But the meet (intersection) of two planes also determines a line. Suppose that we are given two planes; what then is the bivector representing the intersection line of these two planes?

Since by Equation (2.8) the dual of a plane is a point, dualizing our formula for the join of two points in the plane model, we should guess that

$$\pi \vee \rho = \rho^{\perp} \wedge \pi^{\perp} \tag{2.15}$$

is the meet of two planes π, ρ in the point model.

Theorem 2.9

The bivector $l = \rho^{\perp} \wedge \pi^{\perp}$ *represents the intersection line of the planes π and ρ.*

Proof: Consider two planes

$$\pi = p_0 F_0 + p_1 F_1 + p_2 F_2 + p_3 F_3 = p_0 F_0 + v_{\pi}$$

$$\rho = q_0 F_0 + q_1 F_1 + q_2 F_2 + q_3 F_3 = q_0 F_0 + v_{\rho}$$

We can always normalize the planes π and ρ so that $p_0, q_0 = 0$ or 1. To simplify our computations, we shall assume that $p_0 = q_0 = 1$. The cases where one or both of p_0, q_0 are zero can be handled similarly. By Equation (2.8)

$$\pi^{\perp} = -(f_0 + p_1 f_1 + p_2 f_2 + p_3 f_3) = -f_0 + v_{\pi}^{\perp}$$

$$\rho^{\perp} = -(f_0 + q_1 f_1 + q_2 f_2 + q_3 f_3) = -f_0 + v_{\rho}^{\perp}$$

$$l = \rho^{\perp} \wedge \pi^{\perp} = (p_1 - q_1) f_{01} + (p_2 - q_2) f_{02} + (p_3 - q_3) f_{03}$$

$$+ (q_3 p_2 - q_2 p_3) f_{32} + (q_1 p_3 - q_3 p_1) f_{13} + (q_2 p_1 - q_1 p_2) f_{21}$$

Split the bivector l into the sum two bivectors

$$\tilde{v} = l_{32} f_{32} + l_{13} f_{13} + l_{21} f_{21}$$

$$= (q_3 p_2 - q_2 p_3) f_{32} + (q_1 p_3 - q_3 p_1) f_{13} + (q_2 p_1 - q_1 p_2) f_{21}$$

$$\tilde{u} = l_{01} f_{01} + l_{02} f_{02} + l_{03} f_{03} = (p_1 - q_1) f_{01} + (p_2 - q_2) f_{02} + (p_3 - q_3) f_{03}$$

$$l = \tilde{v} + \tilde{u}$$

and set

$$v = l_{32} f_1 + l_{13} f_2 + l_{21} f_3$$

$$= (p_2 q_3 - p_3 q_2) f_1 + (p_3 q_1 - p_1 q_3) f_2 + (p_1 q_2 - p_2 q_1) f_3 = -(f_0 \wedge l)^{\perp}$$

$$u = l_{01} f_1 + l_{02} f_2 + l_{03} f_3 = (p_1 - q_1) f_1 + (p_2 - q_2) f_2 + (p_3 - q_3) f_3 = (f_0 \wedge l^{\perp})^{\perp}.$$

Then

$$u = v_{\rho}^{\perp} - v_{\pi}^{\perp}$$

$$v = v_{\pi}^{\perp} \times v_{\rho}^{\perp}.$$

Now we need to show three things:

i. The bivector $l = \rho^{\perp} \wedge \pi^{\perp}$ represents a line.
ii. The line represented by the bivector l is parallel to the intersection of the two planes π and ρ.
iii. There is a point p on the line l that is also on the intersection of the two planes π and ρ.

We shall prove each statement in turn.

i. It is straightforward to verify (see Exercise 3) that

$$l_{01}l_{32} + l_{02}l_{13} + l_{03}\; l_{21} = u \cdot v = 0.$$

Therefore by Section 2.4.2, the bivector $l = \rho^{\perp} \wedge \pi^{\perp}$ represents a line.

ii. By Corollary 2.7, the direction of the line l is given by the vector $v = -(f_0 \wedge l)^{\perp}$. But $v = v_{\pi}^{\perp} \times v_{\rho}^{\perp}$, which is a vector perpendicular to the normal vectors of the planes π and ρ, and therefore parallel to the direction of the intersection line of π and ρ.

iii. It remains then to show that there is a point p on the line $l = \rho^{\perp} \wedge \pi^{\perp}$ that is also on the intersection of the two planes π and ρ. By Corollary 2.7, the point

$$p = f_0 + \frac{u \times v}{|v|^2}$$

lies on the line l. We shall show that the point p also lies on the two planes π and ρ. Recall that by Theorem 2.4 the point p lies on the plane π if and only if $p \wedge \pi = 0$. Now

$$p \wedge \pi = \left(f_0 + \frac{u \times v}{|v|^2} \right) \wedge (F_0 + v_{\pi}) = f_{0123} + \frac{u \times v}{|v|^2} \wedge v_{\pi}$$

$$\frac{u \times v}{|v|^2} \wedge v_{\pi} = -\left(\frac{u \times v}{|v|^2} \cdot v_{\pi}^{\perp} \right) f_{0123} \qquad \text{(see Exercise 4)}$$

$$= -\frac{\left(\left(v_{\rho}^{\perp} - v_{\pi}^{\perp} \right) \times \left(v_{\pi}^{\perp} \times v_{\rho}^{\perp} \right) \right) \cdot v_{\pi}^{\perp}}{\left| v_{\pi}^{\perp} \times v_{\rho}^{\perp} \right|^2} f_{0123}$$

$$\left(\left(v_\pi^\perp \times v_\rho^\perp\right)\times\left(v_\rho^\perp - v_\pi^\perp\right)\right)\cdot v_\pi^\perp = \det\left(v_\pi^\perp \times v_\rho^\perp \quad v_\rho^\perp - v_\pi^\perp \quad v_\pi^\perp\right)$$

$$= \det\left(v_\pi^\perp \times v_\rho^\perp \quad v_\rho^\perp \quad v_\pi^\perp\right)$$

$$\det\left(v_\pi^\perp \times v_\rho^\perp \quad v_\rho^\perp \quad v_\pi^\perp\right) = \left(v_\pi^\perp \times v_\rho^\perp\right)\cdot\left(v_\rho^\perp \times v_\pi^\perp\right) = -\left|v_\pi^\perp \times v_\rho^\perp\right|^2.$$

Putting all these formulas together yields

$$p \wedge \pi = f_{0123} - \frac{\left|v_\pi^\perp \times v_\rho^\perp\right|^2}{\left|v_\pi^\perp \times v_\rho^\perp\right|^2} f_{0123} = 0,$$

so the point p lies on the plane π. An analogous argument shows that the point p also lies on the plane ρ (see Exercise 10). Therefore, the point p on the line l also lies on the intersection of the planes π and ρ. ◊

Exercises

1. Show that the meet $\pi \vee \rho$ is an antisymmetric, bilinear function. That is, show that for three planes π, ρ, τ:

   a. $\pi \vee \rho = -\rho \vee \pi$

   b. $\left(a\pi + b\rho\right)\vee\tau = a\left(\pi \vee \tau\right)+b\left(\rho \vee \tau\right)$

   c. $\pi \vee \left(c\rho + d\tau\right)=c\left(\pi \vee \rho\right)+d\left(\pi \vee \tau\right).$

2. Show that up to constant multiples, the line represented by the meet of two planes is independent of the choice of the two planes whose intersection is the line. (Hint: Apply Exercise 1.)

3. Consider the two vectors u, v defined in the proof of Theorem 2.9. Show that $u \cdot v = 0$:

   a. by direct computation

   b. by using standard properties of the cross product.

4. Let u be a vector (grade 1) and let v be a normal vector (grade 3). Show that

$$u \wedge v = -\left(u \cdot v^\perp\right) f_{0123}.$$

5. Show that:

   a. $p^{\perp} \wedge \pi^{\perp} = p \wedge \pi$ for any point p and any plane π.

   b. $\left(\pi_1^{\perp} \wedge \pi_2^{\perp} \wedge \pi_3^{\perp} \right)^{\perp}$ is the intersection point of the three independent planes π_1, π_2, π_3.

6. Let l be a line and let π be a plane. Using Exercise 5, show that:

   a. $p = \left(l \wedge \pi^{\perp} \right)^{\perp}$ is the intersection point of l and π if l and π intersect.

   b. $v_l = \left(l \wedge \pi^{\perp} \right)^{\perp}$ if the line l in the direction of the vector v_l is parallel to the plane π.

   c. $0 = l \wedge \pi^{\perp}$ if the line l lies on the plane π.

7. Show that two lines l_1, l_2 are skew if and only if $l_1 \wedge l_2 \neq 0$. (Hint: Apply Exercise 5.)

8. Consider a line represented by the bivector

$$l = l_{32} f_{32} + l_{13} f_{13} + l_{21} f_{21} + l_{01} f_{01} + l_{02} f_{02} + l_{03} f_{03}.$$

   As in the proof of Theorem 2.9, let

$$v = l_{32} f_1 + l_{13} f_2 + l_{21} f_3$$

$$u = l_{01} f_1 + l_{02} f_2 + l_{03} f_3.$$

   Show that

   a. $l = f_0 u - F_0 v$.

   b. $\left(v, u^{\perp} \right)$ are equivalent to the dual Plucker coordinates of the line l.

9. Consider the two lines

$$\tilde{v} = \left(b_3 a_2 - b_2 a_3 \right) f_{32} + \left(b_1 a_3 - b_3 a_1 \right) f_{13} + \left(b_2 a_1 - b_1 a_2 \right) f_{21}$$

$$\tilde{u} = \left(b_1 a_0 - b_0 a_1 \right) f_{10} + \left(b_2 a_0 - b_0 a_2 \right) f_{20} + \left(b_3 a_0 - b_0 a_3 \right) f_{30}$$

   Show that:

   a. The line $\tilde{v}$ is the intersection of two planes passing through the origin.

   b. The line $\tilde{u}$ is a line at infinity.

10. Complete the proof of Theorem 2.9 by showing that the point p also lies on the plane ρ.

11. In the proof of Theorem 2.9, we tacitly assumed that the planes π and ρ are not parallel. Show that if the planes π and ρ are parallel, then $l = \rho^{\perp} \wedge \pi^{\perp}$ is a line at infinity.

12. Given a bivector l representing a line, show how to find two planes that intersect in the line l. That is, show how to construct two planes π, ρ so that $l = \rho^{\perp} \wedge \pi^{\perp}$. (Hint: Use Lemma 2.6.)

---

3 Transformations: Rotors and Versors

In Part I we show how to compute rotation, translation, and reflection on vectors, points, and planes in the space of dual quaternions. In Sections 2.1 and 2.2, we show that multiplication by f_0 maps vectors, points, and planes in the space of dual quaternions into the corresponding vectors, points, and planes in the point model of Clifford algebra. *Therefore our approach to deriving the formulas for these transformations in the point model for dual quaternions will be to use the formulas for rotation, translation, and reflection on vectors, points, and planes in the space of dual quaternions together with multiplication by f_0 to generate the corresponding formulas for rotation, translation, and reflection on vectors, points, and planes in the point model of Clifford algebra.* We shall see that with this approach, the proofs of these transformation formulas simplify considerably; moreover, the proofs of the formulas for these transformations on lines follow readily from the formulas for these transformations on points, since by Sections 2.4.1 and 2.4.2 the bivectors representing lines are the duals of the wedge products of pairs of points. To derive compact formulas for perspective and pseudo-perspective, we shall use much the same approach along with duality in the point model of Clifford algebra. We shall also see that it is in the derivations of the formulas for these transformations, especially for translation and reflection, that we finally take full advantage of the infinitesimal nature of α.

3.1 Translation

We initiate our study of rigid motions with translations. To proceed, we need the following lemma. Notice that in the proof of this lemma we use the fact that compared to 1, α is infinitesimal.

Lemma 3.1

Let $T = 1 - \frac{1}{2}at$, where $t = f_0 \left(t_1 f_1 + t_2 f_2 + t_3 f_3 \right) \alpha$ and a is a scalar. Then

   a. $T^{-1} = 1 + \frac{1}{2}at$

   b. $Tf_0 = f_0 T^{-1}$.

Proof: Since $f_0^2 = 1 / \alpha$, it follows that

$$\left(1 - \frac{1}{2}at \right) \left(1 + \frac{1}{2}at \right) = 1 + \frac{1}{4}a^2 \left(t_1^2 + t_2^2 + t_3^2 \right) \alpha \to 1$$

since α is infinitesimal. Hence $T^{-1} = \left(1 + \frac{1}{2}at \right)$. Moreover $f_j f_0 = -f_0 f_j$; therefore $tf_0 = -f_0 t$, so $Tf_0 = f_0 T^{-1}$. ◊

Theorem 3.2 (Translation)

Let $t = t_1 f_1 + t_2 f_2 + t_3 f_3$ be a unit vector and let $t = f_0 \alpha t = f_0 \left(t_1 f_1 + t_3 f_3 \right) \alpha$. In addition, let a be a scalar, and set

$$T = 1 - \frac{1}{2}af_0 \alpha t = 1 - \frac{1}{2}at.$$

1. *The sandwiching map*

$$x \mapsto TxT^{-1}$$

- *leaves vectors unchanged*
- *translates points and planes by the distance a in the direction t.*

2. The sandwiching map

$$l \mapsto \left(Tl^\perp T^{-1} \right)^\perp$$

- *translates lines by the distance a in the direction t.*

Proof: Since points have grade 1, we begin with points.

1a. **Points**: To establish this result for points, we shall invoke the correspondence between points in the point model of Clifford algebra and points in the algebra of dual quaternions. Consider then an arbitrary point p and let $p = f_0\tilde{p}$, where $\tilde{p}$ is the dual quaternion corresponding to p. Then by Lemma 3.1, $Tf_0 = f_0 T^{-1}$, so

$$TpT^{-1} = Tf_0\tilde{p}T^{-1} = f_0 T^{-1}\tilde{p}T^{-1}.$$

But since

$$t = f_0\left(t_1 f_1 + t_2 f_2 + t_3 f_3\right)\alpha = \left(t_1 f_{32} + t_2 f_{13} + t_3 f_{21}\right)f_{0123}\alpha \leftrightarrow \left(t_1 i + t_2 j + t_3 k\right)\varepsilon,$$

corresponds to a unit quaternion multiplied by ε, the map $T^{-1} = 1 + \dfrac{1}{2}at$ corresponds to a translation in the space of dual quaternions. Indeed by Part I, Theorem 4.1, the map

$$\tilde{p} \mapsto T^{-1}\tilde{p}T^{-1}$$

translates points $\tilde{p}$ by the distance a in the direction t. Therefore the map

$$p \mapsto TpT^{-1} = (Tf_0\tilde{p}T^{-1}) = f_0(T^{-1}\tilde{p}T^{-1})$$

translates the corresponding points $p = f_0\tilde{p}$ by the distance a in the corresponding direction $\boldsymbol{t} = f_0 t$. For an alternative proof, see Exercise 2.

1b. **Vectors**: The exact same proof works for vectors, using the fact that in the space of dual quaternions vectors are unaffected by translation. For an alternative proof, see Exercise 1.

1c. **Planes**: The proof for planes is analogous to the proof for points, this time invoking the correspondence between planes in the point model of Clifford algebra and planes in the algebra of dual quaternions – see Exercise 4a. For an alternative proof, see Exercise 4b.

2. **Lines**: Consider a line represented by the bivector l. Then by Section 2.4.2, $l = (p_2 \wedge p_1)^{\perp}$, where p_1, p_2 are points on the line l. Moreover since by Equation (2.11) for even grade elements $(r^{\perp})^{\perp} = r$, and since by Chapter 1, Equation (3.7), $T(p_2 \wedge p_1)T^{-1} = T(p_2 T^{-1}) \wedge (Tp_1 T^{-1})$, it follows that

$$\left(Tl^{\perp}T^{-1}\right)^{\perp} = \left(T(p_2 \wedge p_1)T^{-1}\right)^{\perp} = \left((Tp_2T^{-1}) \wedge (Tp_1T^{-1})\right)^{\perp}$$

Now by part 1a, the map $p \mapsto TpT^{-1}$ translates points p by the distance a in the direction t. Therefore the map $\left(Tl^{\perp}T^{-1}\right)^{\perp}$ translates lines l by the distance a in the direction t. $\lozenge$

Notice that like the plane model, the point model reverses the signs of T from dual quaternions and maps one factor of T to T^{-1}. Notice too that for planes if the direction of translation is perpendicular to the normal to the plane, then the points on the plane are translated along the plane, but the plane itself—that is, the trivector representing the plane—does not change. Similarly for lines, if the direction of translation is parallel to the line, then the points on the line are translated along the line, but the line itself—that is, the bivector representing the line—does not change (see Exercise 7). For similar observations for planes in the space of dual quaternions, see Part I, Section 1.4.4.

The expression $T = 1 - \dfrac{1}{2}at$ in Theorem 3.2 that represents translation is called a *rotor* because translation is a rotation in 8-dimensions (see Part I, Section 1.5). The translation rotor can be expressed as an exponential, since $(f_0\alpha)^2 = \alpha \to 0$ implies that $T = e^{-\frac{1}{2}at}$ (see Exercise 9).

Remark 3.3 (Discarding Infinitesimal Terms)

i. *Lemma 3.1 and Theorem 3.2 are the main places in the point model where after a computation we discard terms multiplied by the infinitesimal α in order to finalize our computation and to complete our proof.*

ii. *Composites of translations are another place where infinitesimal terms need to be discarded. Consider two vectors*

$$t = t_1 f_1 + t_2 f_2 + t_3 f_3 \ \ \text{and} \ \ u = u_1 f_1 + u_2 f_2 + u_3 f_3$$

along with the corresponding translation rotors

$$T_t = 1 + \frac{1}{2}f_0\alpha t \ \ \text{and} \ \ T_u = 1 + \frac{1}{2}f_0\alpha u.$$

The composite translation rotor T must be given by

$$T = 1 + \frac{1}{2} f_0 \alpha(t + u) = T_t T_u.$$

But

$$T_t T_u = (1 + \frac{1}{2} f_0 \alpha t)(1 + \frac{1}{2} f_0 \alpha u) = 1 + \frac{1}{2} f_0 \alpha(t + u) - \frac{1}{4} tu\alpha.$$

Evidently then we need to discard the term $\frac{1}{4} tu\alpha$. We can justify this discard in the following manner. Observe that

$$tu = t \cdot u + t \wedge u.$$

Now $t \cdot u$ is a scalar, so the scalar $(t \cdot u)\alpha$ is infinitesimal compared to the scalar 1 and therefore can safely be discarded. Moreover, $t \wedge u$ consists solely of terms of the form constant $\times f_i f_j$ where $i, j \neq 0$. But $f_i f_j \alpha$ is infinitesimal compared to $f_0 f_j \alpha$ for the following reason: $f_i \ll f_0$ because $|f_i| = 1 \ll \infty \approx \frac{1}{a} = |f_0|$, so $f_i f_j \alpha \ll f_0 f_j \alpha$. Therefore we can discard the $f_i f_j$ terms arising from $t \wedge u$. What remains is precisely the formula for the composite translation rotor T.

Discarding infinitesimal terms is hidden in the proof of Theorem 3.2, where we rely only on Lemma 3.1 and the correspondence between points, vectors, and planes in the point model of Clifford algebra and in the algebra of dual quaternions. To see more explicitly how this device of discarding terms multiplied by the infinitesimal α after a computation leads to the correct final result, consider the following two examples.

Examples 3.4 (Translation at the Origin)

Let t be a unit vector and consider the translation rotor $T = 1 - \frac{1}{2} a f_0 \alpha t$. On the origin f_0:

$$T f_0 T^{-1} = \left(1 - \frac{1}{2} a f_0 \alpha t\right) f_0 \left(1 + \frac{1}{2} a f_0 \alpha t\right) = \left(f_0 + \frac{1}{2} at\right)\left(1 + \frac{1}{2} a f_0 \alpha t\right)$$

$$= f_0 + \frac{1}{2} at + \frac{1}{2} at - \frac{1}{4} a^2 f_0 \alpha.$$

Discarding the infinitesimal term $-\frac{1}{4}a^2 f_0\alpha$ yields

$$T f_0 T^{-1} = f_0 + at.$$

Thus we see how the rotor T translates the origin f_0 by the distance a in the direction t.

Examples 3.5 (Translation on the Basis Vectors)

Again let t be a unit vector and consider the translation rotor $T = 1 - \frac{1}{2} a f_0 \alpha t$. On a vector f_j:

$$T f_j T^{-1} = \left(1 - \frac{1}{2} a f_0 \alpha t\right) f_j \left(1 + \frac{1}{2} a f_0 \alpha t\right) = \left(f_j + \frac{1}{2} a t f_0 f_j \alpha\right)\left(1 + \frac{1}{2} a f_0 \alpha t\right)$$

$$= f_j + \frac{1}{2}\left(a t f_0 f_j\right)\alpha - \frac{1}{2}\left(a f_0 f_j t\right)\alpha - \frac{1}{4}(a^2 t f_j t)\alpha$$

The trivector terms cancel and discarding the remaining infinitesimal vector terms yields

$$T f_j T^{-1} = f_j.$$

Thus we see how the rotor T leaves the basis vectors f_j unchanged.

Exercises

1. Prove Theorem 3.2 for arbitrary vectors v by direct computation using Example 3.5.
2. Prove Theorem 3.2 for points by direct computation using Exercise 1 and Example 3.4.
3. Prove by direct computation that the plane at infinity is unaffected by the map in Theorem 3.2—that is, prove that $T F_0 T^{-1} = F_0$. (Hint: Discard infinitesimal terms.)
4. Prove Theorem 3.2 for planes in two ways:
   a. by invoking the correspondence between planes in the point model of Clifford algebra and planes in the algebra of dual quaternions;
   b. by direct computation discarding infinitesimal terms.

5. Prove that Theorem 3.2 is valid for planes if and only if Theorem 3.2 is valid for points by using the fact that a point p lies on a plane π if and only if $p \wedge \pi = 0$. (Hint: Use the fact that translations perpendicular to the normal vector of the plane π do not alter the plane π.)

6. Prove Theorem 3.2 for lines using the fact that the point p lies on the line l if and only if

$$p \wedge l^{\perp} = 0.$$

7. Let $l = l_{32} f_{32} + l_{13} f_{13} + l_{21} f_{21} + l_{01} f_{01} + l_{02} f_{02} + l_{03} f_{03}$ be a bivector representing a line, where $\mathbf{v} = -(f_0 \wedge l)^{\perp} = l_{32} f_1 + l_{13} f_2 + l_{21} f_3$ is a unit vector, and set

$$v = f_0 \alpha \mathbf{v} = f_0 \left(l_{32} f_1 + l_{13} f_2 + l_{21} f_3 \right) \alpha$$

$$T = 1 - \frac{1}{2} f_0 \alpha \mathbf{v} = 1 - \frac{1}{2} v.$$

a. Show that $\left(T l^{\perp} T^{-1} \right)^{\perp} = l$ by appealing to Theorem 3.2 and Corollary 2.7.

b. Give a geometric reason for the result in part a.

c. Verify the result in part a by direct computation.

8. Consider a plane π with unit normal vector v, and set

$$\mathbf{u} = v^{\perp} = \text{a unit vector parallel to } v$$

$$T = 1 + \frac{1}{2} a f_0 \alpha \mathbf{u}.$$

a. Show that $T \pi T^{-1} = \pi - a F_0$ by appealing to Theorem 3.2.

b. Give a geometric reason for the result in part a.

c. Verify the result in part a by direct computation.

9. Let $T = 1 - \frac{1}{2} a t$ be the translation rotor in Theorem 3.2. Show that

$$T = e^{-\frac{1}{2} a t}.$$

(Hint: Use the Taylor expansion for e^x and the infinitesimal property of α.)

3.2 Rotation

We continue our study of rigid motions with rotations. Translations depend on the bivectors f_{01}, f_{01}, f_{03}; rotations depend on the complementary bivectors f_{32}, f_{13}, f_{21}—that is, the quaternions. To proceed, we need the following analogue of Lemma 3.1. Notice that for rotation, unlike translation, we do not need to appeal to the infinitesimal property of α.

Lemma 3.6

Let $R = \cos(\theta/2) + \tilde{v}\sin(\theta/2)$, *where* $\tilde{v} = v_1 f_{32} + v_2 f_{13} + v_3 f_{21}$ *is a unit quaternion. Then*

a. $R^{-1} = R^*$

b. $Rf_0 = f_0 R$.

Proof: Since $\tilde{v}$ is a unit quaternion, $\tilde{v}^2 = -1$. Hence

$$\left(\cos(\theta/2) + \tilde{v}\sin(\theta/2)\right)\left(\cos(\theta/2) - \tilde{v}\sin(\theta/2)\right) = 1.$$

Therefore $R^{-1} = \cos(\theta/2) - \tilde{v}\sin(\theta/2) = R^*$

Moreover, $\tilde{v}f_0 = f_0\tilde{v}$; therefore $Rf_0 = f_0 R$. $\lozenge$

Theorem 3.7 (Rotation about a Line through the Origin)

Let $\tilde{v} = v_1 f_{32} + v_2 f_{13} + v_3 f_{21}$ *be a unit quaternion, and set*

$$R = \cos(\theta/2) + \tilde{v}\,\sin(\theta/2).$$

1. *The sandwiching map*

$$x \mapsto RxR^{-1}$$

rotates vectors, points, and planes by the angle θ around the line through the origin represented by the bivector $\tilde{v}$.

2. *The sandwiching map*

$$l \mapsto \left(Rl^{\perp}R^{-1}\right)^{\perp}$$

rotates lines by the angle θ around the line through the origin represented by the bivector $\tilde{v}$.

Proof: Once again we begin with points.

1a. **Points**: To establish this result for points, we again invoke the correspondence between points in the point model of Clifford algebra and points in the algebra of dual quaternions. Consider then an arbitrary point p and let $p = f_0\tilde{p}$, where $\tilde{p}$ is the dual quaternion corresponding to p. By Lemma 3.6, $R^{-1} = R^*$ and $Rf_0 = f_0R$. Hence by Part I, Theorem 4.1, since $\tilde{v}$ corresponds to a unit quaternion, in the space of dual quaternions the map $\tilde{p} \mapsto R\tilde{p}R^{-1}$ rotates the point $\tilde{p}$ by the angle θ around the vector

$$(v_1 i + v_2 j + v_3 k)\varepsilon \rightarrow \tilde{v}f_{0123}\alpha = (v_1 f_{01} + v_2 f_{02} + v_3 f_{03})\alpha.$$

Therefore the map

$$p \mapsto RpR^{-1} = Rf_0\tilde{p}R^{-1} = f_0\left(R\tilde{p}R^{-1}\right)$$

rotates the corresponding point p by the angle θ around the line through the origin parallel to the corresponding vector

$$f_0\left(\tilde{v}f_{0123}\alpha\right) = \tilde{v}F_0 = v_1 f_1 + v_2 f_2 + v_3 f_3.$$

But by Case 1 in Section 2.4.2, this line is precisely the line represented by the bivector $\tilde{v}$. For an alternative proof, see Exercise 4.

1b. **Vectors**: The exact same proof works as well for vectors—see Exercise 1.

1c. **Planes**: The proof for planes is analogous to the proof for points, this time invoking the correspondence between planes in the point model of Clifford algebra and planes in the algebra of dual quaternions—see Exercise 2.

2. **Lines**: Consider a line represented by the bivector l. Then by Section 2.4.2, $l = (p_2 \wedge p_1)^{\perp}$, where p_1, p_2 are points on the line l. Moreover since by Equation (2.11) for even grade elements $(r^{\perp})^{\perp} = r$, and since by Chapter 1, Equation (3.7), $R(p_2 \wedge p_1)R^{-1} = R(p_2R^{-1}) \wedge (Rp_1R^{-1})$, it follows that

$$\left(Rl^{\perp}R^{-1}\right)^{\perp} = \left(R(p_2 \wedge p_1)R^{-1}\right)^{\perp} = \left((Rp_2R^{-1}) \wedge (Rp_1R^{-1})\right)^{\perp}$$

Now by part 1, the map $p \mapsto RpR^{-1}$ rotates points p by the angle θ around the line through the origin represented by the bivector $\tilde{v}$. Therefore the map $\left(Rl^{\perp}R^{-1}\right)^{\perp}$ rotates lines l by the angle θ around the line through the origin represented by the bivector $\tilde{v}$. ◊

Rigid motions can now be computed by composing rotations and translations using the Clifford product:

Rotation followed by Translation: $x \mapsto T\left(RxR^{-1}\right)T^{-1} = (TR)x(TR)^{-1}$

Translation followed by Rotation: $x \mapsto R\left(TxT^{-1}\right)R^{-1} = (RT)x(RT)^{-1}$.

Thus, the composite of a rotation and a translation can be represented by the Clifford product of the corresponding transformations. Notice the similarity to composing these transformations using dual quaternions, but here the conjugate is replaced by the inverse.

The expression $R = \cos(\theta/2) + \tilde{v}\sin(\theta/2)$ in Theorem 3.7 that represents rotation is also called a *rotor*. Just like the translation rotor, the rotation rotor too can be expressed as an exponential: $R = e^{\frac{1}{2}\tilde{v}\theta}$ (see Exercise 8).

Theorem 3.8 (Rotation around an Arbitrary Line)

Let $l = l_{32} f_{32} + l_{13} f_{13} + l_{21} f_{21} + l_{01} f_{01} + l_{02} f_{02} + l_{03} f_{03}$ be a bivector representing an arbitrary line with unit quaternion $\tilde{v} = l_{32} f_{32} + l_{13} f_{13} + l_{21} f_{21}$. Let $C = f_0 + t_1 f_1 + t_2 f_2 + t_3 f_3$ be a point on the line l, and set

$$v = \tilde{v}F_0 = l_{32} f_1 + l_{13} f_2 + l_{21} f_3$$

$$t = C - f_0 = t_1 f_1 + t_2 f_2 + t_3 f_3$$

$$R = \cos(\theta/2) + \tilde{v}\sin(\theta/2)$$

$$S = R + \sin(\theta/2)(t \wedge v) f_{0123}\alpha$$

Then:

1. *The map $x \mapsto SxS^{-1}$ rotates points, vectors, and planes by the angle θ about the line l.*

2. *The map* $l_* \mapsto \left(Sl_*^{\perp}S^{-1}\right)^{\perp}$ *rotates lines* l_* *by the angle* θ *about the line* l.

Proof:

1. Let

$$t = -tF_0 = t_1 f_{32} + t_2 f_{13} + t_3 f_{21}.$$

$$f_0 \alpha t = (t_1 f_0 f_1 + t_2 f_0 f_2 + t_3 f_0 f_3)\alpha = (t_1 f_{32} + t_2 f_{13} + t_3 f_{21}) f_{0123}\alpha = tf_{0123}\alpha.$$

$$T = 1 - \frac{1}{2} f_0 \alpha t = 1 - \frac{1}{2} tf_{0123}\alpha.$$

Then by Theorem 3.2, T represents translation by the vector t, so T^{-1} represents translation by the vector $-t$. Therefore T^{-1} maps the point C to the origin f_0. Since by Theorem 3.7 the map $x \mapsto RxR^{-1}$ rotates points, vectors, and planes by the angle θ around the line through the origin represented by the bivector $\tilde{v}$, which by Corollary 2.7 is a line through the origin parallel to the line l, it remains only to show that $S = TRT^{-1}$. Now the rest of this proof is almost step by step identical to the proof of Theorem 4.2 in Part I with the wedge product replacing the cross product:

$$TRT^{-1} = \left(1 - \frac{1}{2} tf_{0123}\alpha\right) R \left(1 + \frac{1}{2} tf_{0123}\alpha\right)$$

$$= \left(R - \frac{1}{2} tf_{0123}\alpha R\right) \left(1 + \frac{1}{2} tf_{0123}\alpha\right)$$

$$= R - \frac{1}{2} tRf_{0123}\alpha + \frac{1}{2} Rtf_{0123}\alpha - \frac{1}{4} tf_{0123}\alpha Rtf_{0123}\alpha$$

$$= R + \frac{Rt - tR}{2} f_{0123}\alpha + \frac{1}{4}(tF_0 RtF_0)\alpha \quad \{\text{Discard Infinitesimal Term}\}$$

$$= R + \sin(\theta/2)\frac{\tilde{v}t - t\tilde{v}}{2} f_{0123}\alpha \quad\quad \{R = \cos(\theta/2) + \tilde{v}\sin(\theta/2)\}$$

$$= R + \sin(\theta/2)\frac{tF_0^2\tilde{v} - \tilde{v}F_0^2 t}{2} f_{0123}\alpha \quad \{F_0^2 = -1\}$$

$$= R + \sin(\theta/2)\frac{tv - vt}{2} f_{0123}\alpha$$

$$= R + \sin(\theta/2)(t \wedge v) f_{0123}\alpha$$

2. The proof for lines l_* follows from the fact that $l_* = (p_2 \wedge p_1)^{\perp}$, where p_1, p_2 are points on the line l_*. Now the result for lines follows from the result for points mimicking the proof for lines in Theorem 3.7. ◊

Remark 3.9

Two observations concerning Theorem 3.8:

1. *We can restore the cross product, since* $(t \wedge v) f_{0123} \alpha = (t \times v) f_0 \alpha.$

2. *By Corollary 2.7, we can choose* $C = f_0 + \dfrac{u \times v}{|v|^2}$, *where* $u = l_{01} f_1 + l_{02} f_2 + l_{03} f_3$. *Since* $t = C - f_0$ *and since* v *is a unit vector orthogonal to* u, *we have* $t \times v = (u \times v) \times v = -u.$

Theorem 3.10 (Screw Transformations—Simultaneous Rotation and Translation)

Let $l = l_{32} f_{32} + l_{13} f_{13} + l_{21} f_{21} + l_{01} f_{01} + l_{02} f_{02} + l_{03} f_{03}$ *be a bivector representing an arbitrary line with unit quaternion* $v = l_{32} f_{32} + l_{13} f_{13} + l_{21} f_{21}$. *Let* $C = f_0 + t_1 f_1 + t_2 f_2 + t_3 f_3$ *be a point on the line l, and set*

$$v = vF_0 = l_{32} f_1 + l_{13} f_2 + l_{21} f_3$$

$$t = C - f_0 = t_1 f_1 + t_2 f_2 + t_3 f_3$$

$$R = \cos(\theta/2) + v \sin(\theta/2)$$

$$S = R + \sin(\theta/2)(t \wedge v) f_{0123} \alpha$$

$$M = S + \frac{1}{2} d\big(\sin(\theta/2) - v \cos(\theta/2)\big) f_{0123} \alpha$$

Then:

1. *The map* $x \mapsto MxM^{-1}$ *first rotates points, vectors, and planes by the angle* θ *about the line l and then translates the resulting points, vectors, and planes in the direction* v *parallel to the line l by the distance d.*

2. *The map $l_* \mapsto \left(Ml_*^{\perp} M^{-1} \right)^{\perp}$ first rotates lines l_* by the angle θ about the line l and then translates the resulting lines in the direction v parallel to the line l by the distance d.*

Proof:

1. By Theorem 3.8 the map $x \mapsto SxS^{-1}$ represents rotation by the angle θ about the line l. Let

$$T = 1 - \frac{d}{2} f_0 \alpha v.$$

By Theorem 3.2, the map $x \mapsto TxT^{-1}$ represents translation by the distance d in the direction of the vector $\mathbf{v} = vF_0$, which is parallel to the direction of the line l. Therefore we need only show that $M = TS$. But $f_0 v = vf_0$, so

$$T = 1 - \frac{d}{2} f_0 \alpha v = 1 - \frac{d}{2} f_0 \alpha v F_0 = 1 - \frac{d}{2} v f_{0123} \alpha.$$

Now we can simply mimic the proof in Part I, Theorem 4.4. Ignoring the infinitesimal term (see Part ii of Remark 3.3):

$$TS = (1 - \frac{d}{2} v f_{0123} \alpha)(R + \sin(\theta/2)(\mathbf{v} \wedge \mathbf{t}) f_{0123} \alpha) = S - \frac{d}{2} v R f_{0123} \alpha.$$

Moreover since v is a unit quaternion $v^2 = -1$, so

$$vR = v\big(\cos(\theta/2) + v\,\sin(\theta/2)\big) = -\sin(\theta/2) + v\,\cos(\theta/2).$$

Hence

$$TS = S + \frac{1}{2} d\big(\sin(\theta/2) - v\cos(\theta/2)\big) f_{0123} \alpha = M.$$

2. Once again the proof for lines l_* follows from the fact that $l_* = (p_2 \wedge p_1)^{\perp}$, where p_1, p_2 are points on the line l_*. The result for lines now follows from the result for points mimicking the proof for lines of Theorem 3.7. $\lozenge$

Remark 3.11

Notice that no infinitesimals appear in Theorem 3.7 to represent and com-
pute the rotation rotor around a line through the origin. Infinitesimals do
appear, however, in Theorems 3.8 and 3.10 because in these theorems the
rotation rotor is composed with a translation rotor.

Exercises

1. Prove Theorem 3.7 for vectors by invoking the correspondence
   between vectors in the point model of Clifford algebra and vectors in the algebra of dual quaternions.

2. Prove Theorem 3.7 for planes by invoking the correspondence
   between planes in the point model of Clifford algebra and planes
   in the algebra of dual quaternions.

3. Prove Theorem 3.7 for the origin f_0 by direct computation using
   the identity $Rf_0 = f_0R$.

4. Prove Theorem 3.7 for points using Exercises 1 and 3.

5. Prove that Theorem 3.7 is valid for planes if and only if Theorem
   3.7 is valid for points by using the fact that a point p lies on a plane
   π if and only if $\pi \wedge p = 0$.

6. Prove Theorem 3.7 for lines using the fact that the point p lies on
   the line l if and only if

$$p \wedge l^{\perp} = 0.$$

7. Prove that the plane at infinity is unaffected by the map in
   Theorem 3.7—that is, prove that

$$RF_0R^{-1} = F_0.$$

8. Let $R = \cos(\theta/2) + \tilde{v}\,\sin(\theta/2)$ be the rotation rotor in Theorem 3.7.
   Show that $R = e^{\frac{1}{2}\tilde{v}\theta}$.
   (Hint: Use the Taylor expansion for e^x and the fact that $\tilde{v}$ is a unit
   quaternion.)

3.3 Reflection

Reflections in the point model are easier than both translations and rotations. Nevertheless, we shall see that just like translations, computations
involving reflections also require discarding infinitesimal terms.

Theorem 3.12 (Reflection in an Arbitrary Plane)

Let π represent an arbitrary plane with a unit normal vector. Then

1. *The sandwiching map*

$$x \mapsto -(\pi x \pi)\alpha$$

reflects points, vectors, and planes in the plane π.

2. *The sandwiching map*

$$l \mapsto -\left(\pi l^{\perp} \pi\right)^{\perp} \alpha$$

reflects lines in the plane π.

Proof: First observe that since π is a plane, there is a dual quaternion $\tilde{\pi}$ such that $\pi = f_0 \tilde{\pi}$.

1a. **Points:** To establish this result for points, we shall once again invoke the correspondence between points in the point model of Clifford algebra and points in the algebra of dual quaternions. Consider then an arbitrary point p and let $p = f_0 \tilde{p}$, where $\tilde{p}$ is the dual quaternion corresponding to p. Then by Lemma 2.1 $\tilde{p} = f_0 \alpha p$ and $\tilde{p}^* = p f_0 \alpha$. Hence

$$(\pi p \pi)\alpha = (f_0 \tilde{\pi})(p)(f_0 \alpha \tilde{\pi}) = f_0 \left(\tilde{\pi} \, \tilde{p}^* \tilde{\pi}\right).$$

Moreover by Part 1, Theorem 4.20, the map

$$\tilde{p} \mapsto -\tilde{\pi} \tilde{p}^* \tilde{\pi}$$

reflects the point $\tilde{p}$ in the plane $\tilde{\pi}$. Therefore the map

$$p \mapsto -(\pi p \pi)\alpha = f_0 \left(-\tilde{\pi} \tilde{p}^* \tilde{\pi}\right)$$

reflects the corresponding point $p = f_0 \tilde{p}$ in the corresponding plane $\pi = f_0 \tilde{\pi}$.

1b. **Vectors:** The proof for vectors v is the same as the proof for points p, this time invoking the correspondence between vectors in the point model of Clifford algebra and vectors in the algebra of dual quaternions—see Exercise 1.

1c. **Planes:** The proof for planes ρ is analogous to the proof for points p, this time invoking the correspondence between planes in the point model of Clifford algebra and planes in the algebra of dual quaternions together with the identity $\tilde{\rho}^* = -\rho f_0 \alpha$ from Lemma 2.1—see Exercise 2. An alternative proof is provided in Exercise 3.

2. **Lines:** Consider a line represented by a bivector l. Then $l = (p_2 \wedge p_1)^\perp$, where p_1, p_2 are points on the line l. Let $\pi = dF_0 + aF_1 + bF_2 + cF_3$. Since π has a unit normal vector, it follows by Equations (1.6) and (1.7) that

$$\pi^2 \alpha = (aF_1 + bF_2 + cF_3 + dF_0)^2 \alpha = -(a^2 + b^2 + c^2) - d^2 \alpha = -1 - d^2 \alpha$$

Discarding the infinitesimal term $d^2 \alpha$, we find that $\pi^{-1} = -\pi \alpha$. Therefore by Chapter 1, Equation (3.7),

$$-\left(\pi l^\perp \pi \alpha\right) = \left(\pi(p_2 \wedge p_1)(-\pi \alpha)\right) = \left(-\pi p_2 \pi \alpha\right) \wedge \left(-\pi p_1 \pi \alpha\right)$$

$$-\left(\pi l^\perp \pi \alpha\right)^\perp = \left(\left(-\pi p_2 \pi \alpha\right) \wedge \left(-\pi p_1 \pi \alpha\right)\right)^\perp.$$

By part 1a, the map $p \mapsto -(\pi p \pi)\alpha$ reflects points p in the plane π. Hence the map $l \mapsto -\left(\pi l^\perp \pi \alpha\right)^\perp$ reflects lines l in the plane π. $\lozenge$

The plane π in Theorem 3.12 that represents reflection is called a *versor*. Unlike rotors, versors cannot be expressed as exponentials.

Three observations are in order here concerning Theorem 3.12. First, notice how the ugly conjugates π^* and p^* that appear in the dual quaternion formulas for reflection (see Part I, Theorem 4.20) disappear from these Clifford algebra formulas for reflection. Second, notice that the representations for reflection in the plane model and the point model are not exactly the same; the point model introduces a minus sign and a factor of α, normalizations not necessary in the plane model (see Chapter 2, Theorem 3.8). Finally, observe that even though α is an infinitesimal, the expression $-(\pi x \pi)\alpha$ does not vanish identically, since it is not added to a non-infinitesimal; rather terms with this α factor either cancel during computations or can be discarded after computations. Let's look at some examples.

Examples 3.13 (Reflection in the Plane $F_3 - F_0$)

The trivector $F_3 - F_0$ represents the plane with the equation $z - 1 = 0$, a plane parallel to the xy-plane ($z = 0$).

    i. The origin f_0:

$$-(F_3 - F_0)f_0(F_3 - F_0)\alpha = -f_0(F_3 + F_0)(F_3 - F_0)\alpha$$
$$= -f_0\left(F_3^2 + 2F_0F_3 - F_0^2\right)\alpha$$
$$= f_0 + 2f_3 + f_0\alpha.$$

Discarding the infinitesimal term $f_0\alpha$, we find that the origin f_0 is reflected to the point $f_0 + 2f_3$.

    ii. The vector f_3 perpendicular to the xy-plane:

$$-(F_3 - F_0)f_3(F_3 - F_0)\alpha = f_3(F_3 + F_0)(F_3 - F_0)\alpha$$
$$= f_3\left(F_3^2 + 2F_0F_3 - F_0^2\right)\alpha$$
$$= -f_3 + 2f_0\alpha + f_3\alpha.$$

Discarding the infinitesimal terms $2f_0\alpha + f_3\alpha$, we find that the vector f_3 perpendicular to the xy -plane is reflected in the plane $F_3 - F_0$ to the vector $-f_3$.

    iii. The vector f_1 perpendicular to the yz-plane:

$$-(F_3 - F_0)f_1(F_3 - F_0)\alpha = -f_1(F_3 - F_0)(F_3 - F_0)\alpha$$
$$= -f_1\left(F_3^2 - F_0^2\right)\alpha$$
$$= f_1 - f_1\alpha.$$

Discarding the *infinitesimal* term $-f_1\alpha$, we find that the vector f_1 perpendicular to the yz-plane is not affected by reflection in the plane $F_3 - F_0$.

    iv. The plane $F_3 + F_0$ parallel to the plane $F_3 - F_0$:

$$-(F_3 - F_0)(F_3 + F_0)(F_3 - F_0)\alpha$$
$$= -(F_3 - F_0)F_3(F_3 - F_0)\alpha - (F_3 - F_0)F_0(F_3 - F_0)\alpha$$

Now using Equations (1.6) and (1.7), we compute each term in turn:

$$-(F_3 - F_0)F_3(F_3 - F_0)\alpha = -F_3(F_3 + F_0)(F_3 - F_0)\alpha$$

$$= -F_3\left(F_3^2\alpha - F_0^2\alpha + 2F_0F_3\alpha\right)$$

$$= -F_3(-1 + \alpha + 2F_0F_3\alpha)$$

$$= F_3 - F_3\alpha - 2F_0 \rightarrow F_3 - 2F_0$$

$$-(F_3 - F_0)F_0(F_3 - F_0)\alpha = F_0(F_3 + F_0)(F_3 - F_0)\alpha$$

$$= F_0\left(F_3^2\alpha - F_0^2\alpha + 2F_0F_3\alpha\right)$$

$$= F_0(-1 + \alpha + 2F_0F_3\alpha)$$

$$= -F_0 + F_0\alpha - 2F_3\alpha \rightarrow -F_0$$

Thus adding and discarding the infinitesimal terms, we find that the plane $F_3 + F_0$ $(z + 1 = 0)$ parallel to the plane $F_3 - F_0$ $(z - 1 = 0)$ is reflected to the plane $F_3 - 3F_0$ $(z - 3 = 0)$. Notice too we have also shown that the plane F_3 $(z = 0)$ is reflected to the plane $F_3 - 2F_0$ $(z - 2 = 0)$.

Exercises

1. Prove Theorem 3.12 for vectors by invoking the correspondence between vectors in the point model of Clifford algebra and vectors in the algebra of dual quaternions.

2. Prove Theorem 3.12 for planes by invoking the correspondence between planes in the point model of Clifford algebra and planes in the algebra of dual quaternions and applying the identities for planes and their conjugates in Lemma 2.1.

3. Prove that Theorem 3.12 is valid for planes if and only if Theorem 3.12 is valid for points by using the fact that the point p lies on the plane π if and only if $p \wedge \pi = 0$.

4. Prove Theorem 3.12 for lines using the fact that the point p lies on the line l if and only if

$$p \wedge l^{\perp} = 0.$$

5. Let π be a plane with a unit normal vector, and let v be an arbitrary vector.
   a. Show that $(-\pi v \pi \alpha)^{\perp} = \pi v^{\perp}\pi\alpha$. (Hint: See Section 2.3.1, Exercise 1.)

b. Conclude that Theorem 3.12 is valid for vectors if and only if Theorem 3.12 is valid for normal vectors. (Hint: π and $-\pi$ represent the same plane.)

6. Let π represent a plane through the origin with a unit normal vector. Show that:

a. The sandwiching map

$$x \mapsto -\pi^{\perp} x \pi^{\perp}$$

reflects points, vectors, and planes in the plane π.

b. The sandwiching map

$$-\left(\pi^{\perp} l^{\perp} \pi^{\perp}\right)^{\perp}$$

reflects lines in the plane π.

7. Let π represent a plane with a unit normal vector, and let $\hat{\pi} = f_{0123} \alpha \pi$. Let x denote a point, vector, or plane. Show that:

a. If the plane π passes through the origin, then
   i. $\hat{\pi} = \pi^{\perp}$
   ii. $-\hat{\pi} x \hat{\pi} = -\pi^{\perp} x \pi^{\perp}$.

b. If the plane π does not pass through the origin, then $-\hat{\pi} x \hat{\pi} = -(\pi x \pi) \alpha$.

(Hint: Mimic the proof of Theorem 3.12.)

c. Conclude from Exercise 6 and Theorem 3.12 that in all cases the sandwiching map $x \mapsto - \hat{\pi} x \hat{\pi}$ reflects points, vectors, and planes in the plane π.

3.4 Perspective and Pseudo-Perspective

Perspective and pseudo-perspective can be computed by using the translation rotor together with duality. To proceed, we need the following lemma relating duality in the space of dual quaternions to duality in the point model.

Lemma 3.14

Let x be a mass-point or a vector, and let $\tilde{x} = f_0 \alpha x$ be the dual quaternion corresponding to x. Then $x^{\perp} = f_0 \tilde{x}^{\#}$.

Proof: We shall prove this result for mass-points; the proof for vectors is similar, just set the mass to zero. Consider then a mass-point p and the corresponding dual quaternion $\tilde{p}$:

$$p = p_0 f_0 + p_1 f_1 + p_2 f_2 + p_3 f_3 = f_0 \tilde{p}$$

$$\tilde{p} = f_0 \alpha p = p_0 + (p_1 f_{01} + p_2 f_{02} + p_3 f_{03})\alpha = p_0 + (p_1 f_{32} + p_2 f_{13} + p_3 f_{21}) f_{0123}\alpha.$$

Now by definition

$$p^\perp = p_0 F_0 + p_1 F_1 + p_2 F_2 + p_3 F_3$$

$$\tilde{p}^\# = p_0 f_{0123}\alpha + p_1 f_{32} + p_2 f_{13} + p_3 f_{21}$$
$$= f_0 \alpha(p_0 F_0 + p_1 F_1 + p_2 F_2 + p_3 F_3) = f_0 \alpha p^\perp.$$

Therefore $p^\perp = f_0 \tilde{p}^\#$. $\Diamond$

Theorem 3.15 (Perspective Projection from the Origin)

Let

- $E = f_0 = $ *eye point*

- $t = t_1 f_1 + t_2 f_2 + t_3 f_3 = $ *a unit vector*

- $t = f_0(t_1 f_1 + t_2 f_2 + t_3 f_3)\alpha = f_0 \alpha t$

- $\pi = F_0 - \dfrac{1}{d} t^\perp = $ *the perspective plane at a distance d from the eye point*
$E = f_0$ *perpendicular to the unit normal vector* $t^\perp$

- $T = 1 + \dfrac{1}{2d} t = $ *a translation rotor*

Then $\left(T(P - E)^\perp T^{-1}\right)^\perp$ *is a mass-point* (mP', m), *where*

- *the point* P' *is located at the perspective projection of the point P from the eye point* $E = f_0$ *onto the plane* π;

- the mass $m = s/d$ is equal to the distance s of the point P from the plane through the eye point $E = f_0$ perpendicular to the unit normal vector $t^\perp$ divided by the fixed distance d from the eye point E to the projection plane π.

Proof: By Theorem 7.2 in Part I, in the space of dual quaternions the mass-point representing perspective projection is given by the dual quaternion $\left(T^{-1}(\widetilde{P-E})^{\#}T^{-1}\right)^{\#}$. To convert from dual quaternions to the point model of Clifford algebra, we multiply the dual quaternion terms by f_0. Hence in the point model of Clifford algebra this formula becomes

$$\left(T^{-1}(\widetilde{P-E})^{\#}T^{-1}\right)^{\#} \rightarrow f_0\left(f_0 T^{-1}(\widetilde{P-E})^{\#}T^{-1}\right)^{\#}$$

Now by Lemma 3.14, $p^\perp = f_0\tilde{p}^{\#}$, and by Lemma 3.1, $f_0 T^{-1} = Tf_0$. Therefore

$$f_0\left(f_0 T^{-1}(\widetilde{P-E})^{\#}T^{-1}\right)^{\#} = f_0\left(Tf_0(\widetilde{P-E})^{\#}T^{-1}\right)^{\#} = \left(T(P-E)^\perp T^{-1}\right)^\perp . \lozenge$$

Our next result shows that on vectors the rotation rotor commutes with duality. Therefore we shall see that we can compose rotation with perspective and pseudo-perspective in the same way that we can compose rotation with translation—by multiplying the transformations.

Lemma 3.16

Let

$\tilde{v} = v_1 f_{32} + v_2 f_{13} + v_3 f_{21} = $ *a unit quaternion*

$R = \cos(\theta/2) + \tilde{v}\,\sin(\theta/2) = $ *a rotation by the angle θ around the line through the origin represented by the bivector $\tilde{v}$*

Then for any vector w

$$\left(RwR^{-1}\right)^\perp = Rw^\perp R^{-1}.$$

Thus on vectors rotation around a line through the origin commutes with duality.

Proof: Observe that

$$w = w_1 f_1 + w_2 f_2 + w_3 f_3 = (w_1 F_1 + w_2 F_2 + w_3 F_3) f_{0123} \alpha = w^{\perp} f_{0123} \alpha$$

Therefore

$$R w R^{-1} = R \left(w^{\perp} f_{0123} \alpha \right) R^{-1} = \left(R w^{\perp} R^{-1} \right) f_{0123} \alpha = -\left(R w^{\perp} R^{-1} \right)^{\perp}$$

Now recall from Equation (2.12) that if grade x is odd, then $\left(x^{\perp} \right)^{\perp} = -x$. Therefore taking the dual of both sides yields

$$\left(R w R^{-1} \right)^{\perp} = R w^{\perp} R^{-1}. \lozenge$$

Since rotation and duality commute, we can compose rotation and perspective projection in the same way that we compose rotation and translation—by taking the product of their representations as dual quaternions. Indeed by Lemma 3.16

Rotation followed by perspective

$$P \mapsto \left(T \left(R(P-E) R^{-1} \right)^{\perp} T^{-1} \right)^{\perp} = \left((TR)(P-E)^{\perp} (TR)^{-1} \right)^{\perp}$$

Perspective followed by rotation

$$P \mapsto R \left(T(P-E)^{\perp} T^{-1} \right)^{\perp} R^{-1} = \left((RT)(P-E)^{\perp} (RT)^{-1} \right)^{\perp}.$$

Note that here the rotation R and the perspective map T are both with respect to the eye point at the origin—that is, the rotation is around a line through the origin.

Theorem 3.17 (Pseudo-Perspective)

Let

- $E = f_0 - d\mathbf{t} = $ *eye point*

- $\mathbf{t} = t_1 f_1 + t_2 f_2 + t_3 f_3 = $ *a unit vector*

- $t = f_0\big(t_1 f_1 + t_2 f_2 + t_3 f_3\big)\alpha = f_0\alpha t$
- $T = 1 + \dfrac{1}{2d}t = $ a translation rotor

Then the map $P \mapsto \big(TP^{\perp}T^{-1}\big)^{\perp}$ transforms the eye point E to the point at infinity in the direction t and transforms a viewing frustum to a rectangular box.

Proof: By Theorem 7.3 in Part I, in the space of dual quaternions pseudo-perspective is given by the map

$$\tilde{P} \mapsto \big(T^{-1}\tilde{P}^{\#}T^{-1}\big)^{\#}.$$

Again to convert from dual quaternions to the point model of Clifford algebra, we multiply the dual quaternion terms by f_0. Therefore in the point model of Clifford algebra this formula becomes

$$P \mapsto f_0\big(f_0(T^{-1}\tilde{P}^{\#}T^{-1})\big)^{\#}.$$

Now by Lemma 3.14, $p^{\perp} = f_0\tilde{p}^{\#}$, and by Lemma 3.1, $f_0 T^{-1} = T f_0$. Therefore

$$P \mapsto f_0\big(f_0(T^{-1}\tilde{P}^{\#}T^{-1})\big)^{\#} = f_0\big(T f_0\tilde{P}^{\#}T^{-1}\big)^{\#} = \big(TP^{\perp}T^{-1}\big)^{\perp}. \Diamond$$

In our theorems for perspective and pseudo-perspective, we have located the eye point E at a special canonical position for which the results are especially easy to compute. To deal with arbitrary eye points E', we can simply translate the entire scene—points and planes—by the vector $v = E - E'$ before performing perspective or pseudo-perspective and then after performing perspective or pseudo-perspective with the eye point in canonical position, translate the result by $-v = E' - E$ to move the result back to the required position. For perspective, the vector $P - E$ is unaffected by translation, so we need to perform only the final translation and can dispense altogether with the initial translation.

Exercises

1. Here we provide a more direct, computational proof of Theorem 3.15. Let $P = f_0 + v_P$. Using the notation of Theorem 3.15, show that:

a. $T(P-E)^{\perp}T^{-1} = v_p^{\perp} + \dfrac{v_p \cdot \mathbf{t}}{d} F_0$

     (Hint: First show that

$$\frac{1}{d}\frac{tv_p^{\perp} - v_p^{\perp}t}{2} = -\frac{F_0\alpha}{d}\left(\frac{\mathbf{t}^{\perp}v_p^{\perp} + v_p^{\perp}\mathbf{t}^{\perp}}{2}\right) = \frac{v_p \cdot \mathbf{t}}{d}F_0.)$$

b. $\left(T(P-E)^{\perp}T^{-1}\right)^{\perp} \equiv f_0 + \left(\dfrac{d}{v_p \cdot \mathbf{t}}\right)v_p$

c. $\left(T(P-E)^{\perp}T^{-1}\right)^{\perp} \wedge \pi = \dfrac{1}{d}\left((v_p \cdot \mathbf{t})f_{0123} - v_p \wedge \mathbf{t}^{\perp}\right)$

d. $\left(v_p^{\perp} \cdot \mathbf{t}^{\perp}\right)f_{0123} = v_P \wedge \mathbf{t}^{\perp}$

Conclude that $\left(T(P-E)^{\perp}T^{-1}\right)^{\perp}$ is a mass-point (mP', m), where:

e.   the point P' is located at the perspective projection of the point P from the eye point $E = f_0$ onto the plane π;

f.   the mass $m = s/d$ is equal to the distance s of the point P from the plane through the eye point $E = f_0$ perpendicular to the unit normal vector $\mathbf{t}^{\perp}$ divided by the fixed distance d from the eye point E to the perspective plane π.

2. Here we give a more direct computational proof of Theorem 3.17. Let $\mathbf{t}$ be the normal vector to the perspective plane and let $\mathbf{t}_{\perp}$ be any vector perpendicular to $\mathbf{t}$. Using the notation of Theorem 3.17, show that:

   a. $\left(Td\mathbf{t}^{\perp}T^{-1}\right)^{\perp} = f_0 + d\mathbf{t}$ (Hint: First show that $\dfrac{\mathbf{t}\mathbf{t}^{\perp} - \mathbf{t}^{\perp}\mathbf{t}}{2} = f_{0123}$.)

   b. $\left(TE^{\perp}T^{-1}\right)^{\perp} = d\mathbf{t}$ (Hint: Discard infinitesimal terms.)

   c. $\left(T\mathbf{t}_{\perp}^{\perp}T^{-1}\right)^{\perp} = -\mathbf{t}_{\perp}$ (Hint: First show that $\dfrac{\mathbf{t}\mathbf{t}_{\perp}^{\perp} - \mathbf{t}_{\perp}^{\perp}\mathbf{t}}{2} = 0$.)

   d.   Conclude that the map $P \mapsto \left(TP^{\perp}T^{-1}\right)^{\perp}$ transforms the eye point E to the point at infinity in the direction $\mathbf{t}$ and transforms a viewing frustum to a rectangular box.

3. Let P be an arbitrary point, and suppose that the eye point $E = f_0$.

   a.   Show that $l = \left((P-E) \wedge E\right)^{\perp}$ is a bivector representing the line through the eye point E parallel to the vector $P - E$.

   b.   Show how to find the intersection of the line l with the plane π. (Hint: See Section 2.4.4, Exercise 6.)

   c.   Use part b to find the perspective projection of the point P on the plane π.

4. Let π be a plane with a unit normal vector, and let v be an arbitrary vector.

   a. Show that $(-\pi v \pi \alpha)^{\perp} = (\pi v^{\perp} \pi) \alpha$.

   b. Conclude that on vectors reflection in a plane commutes with duality.

5. In this exercise we show how to compute perspective projection using the rotation rotor in the point model. Let

   - $v =$ a unit quaternion
   - $u = F_0 v =$ a unit vector
   - $E = f_0 - u =$ eye point
   - $\pi = u^{\perp} =$ perspective plane through the origin perpendicular to the unit vector u
   - $R = \cos(\pi/4) - \sin(\pi/4)v =$ rotation by $-\pi/2$ around the line through the origin parallel to the unit vector u

   Show that $R(P - E)R$ is a mass-point (mP', m), where:

   - the point P' is located at the perspective projection of the point P from the eye point E onto the plane π;
   - the mass m is equal to the distance d of the point P from the plane through the eye point E perpendicular to the unit vector u.

   (Hint: Apply [9, Chapter 6, Theorem 6.8].)

6. Repeat Exercise 5 this time with

   - $v =$ a unit quaternion
   - $u = F_0 v =$ a unit vector
   - $E = f_0 + (\cot(\theta) - \csc(\theta))u =$ eye point
   - $\pi = -\cot(\theta)F_0 + u^{\perp} =$ perspective plane perpendicular to the unit vector u at a distance $\csc(\theta)$ from the eye point E
   - $R = \cos(\theta/2) - \sin(\theta/2)v =$ rotation by $-\theta$ around the line through the origin parallel to the unit vector u

   Show that $R(P - E)R$ is a mass-point (mP', m), where:

   - the point P' is located at the perspective projection of the point P from the eye point E onto the plane π;
   - the mass $m = d\sin(\theta)$, where d is the distance of the point P from the plane through the eye point E perpendicular to the unit vector u.

   (Hint: Apply [9, Chapter 6, Theorem 6.9].)

4 Insights

Here are seven of the main insights that guided our intuition for investigating and deriving the properties of the point model of Clifford algebra for dual quaternions.

1. *Quaternions and dual quaternions reside inside the point model*
   - $f_{32} \leftrightarrow i \quad f_{13} \leftrightarrow j \quad f_{21} \leftrightarrow k \quad f_{0123}\alpha \leftrightarrow \varepsilon$

2. *The origin f_0 maps from the dual quaternion to the point model representation of geometry*
   - $x \mapsto f_0 x$

3. *Lines are represented as the join of two points using Plucker coordinates*

4. *Duality is used to define and compute*
   - *Join of two points*
   - *Meet of two planes*
   - *Translation, rotation, and reflection on lines*
   - *Perspective and pseudo-perspective*

5. *Infinitesimals*
   Infinitesimals are used almost exclusively to compute translations and reflections.
   Infinitesimals are also used in the formulas for perspective and pseudo-perspective where the translation rotor is applied.

6. *Representations*
   - *Rotors (operators) have even grade; points and planes (operands) have odd grade.*

7. *The point model is a reciprocal version of the plane model*
   - *Algebra: $g_0^2 = 0 \rightarrow f_0^2 = 1/\alpha \approx \infty$*
   - Geometry: Representations of geometry in the point model are reciprocal versions of representations of geometry in the plane model.
     To get from the plane model to the point model:
     - for points, vectors, and planes: replace g's by f's, and exchange lower case and upper case letters;
     - for lines: replace each factor in the meet or the join by its dual.

- Duality: to get from the plane model to the point model, just replace g's by f's.
- Transformations:
  - Rotors: to get from the plane model to the point model, just replace g's by f's, and introduce a factor of α in the translation rotor.
  - Versors: to get from the plane model to the point model introduce a minus sign and a factor of α, and separate out the line as a special case.

5 Formulas

We provide a summary here for easy future reference of the main formulas we have encountered for the point model of dual quaternions.

5.1 Algebra

Notation

$$f_{ij} = f_i f_j, \quad i \neq j.$$

$$F_0 = f_1 f_2 f_3, \quad F_1 = f_0 f_3 f_2, \quad F_2 = f_0 f_1 f_3, \quad F_3 = f_0 f_2 f_1.$$

$$f_{0123} = f_0 f_1 f_2 f_3.$$

Identities

$$f_0^2 = 1/\alpha, \; f_1^2 = f_2^2 = f_3^2 = 1$$

$$f_i \cdot f_j = 0 \qquad i \neq j$$

$$f_i f_j = -f_j f_i \quad i \neq j$$

$$f_i F_i = f_{0123} = -F_i f_i, \quad i = 0,1,2,3$$

$$f_i F_0 = f_{jk} = F_0 f_i, \quad ijk \text{ an even permutation of } 123$$

$$f_j F_i = sgn(ijk) f_{0k} = F_i f_j, \quad ijk \text{ a permutation of } 123$$

$$F_i F_j = -F_j F_i \qquad j \neq i$$

$$F_j^2 \alpha = F_0^2 = -1. \ j \neq 0$$

Products $(u, v \text{ grade } 1)$

$$uv = u \cdot v + u \wedge v$$

$$u \cdot v = \frac{uv + vu}{2}$$

$$u \wedge v = \frac{uv - vu}{2}$$

$$u \times v = -F_0 (u \wedge v)$$

Wedge Product $(\text{grade}(U) = c, \text{grade}(V) = d)$

$$U \wedge V = \frac{UV + (-1)^{cd} VU}{2}$$

$$s(U \wedge V)s^{-1} = \left(sUs^{-1} \right) \wedge \left(sVs^{-1} \right)$$

5.2 Geometry

Representations

	Clifford Algebra (Point Model)	Coordinate Representation
Points	$p = f_0 + p_1 f_1 + p_2 f_2 + p_3 f_3$	$p = (p_1, p_2, p_3)$
Vectors	$v = v_1 f_1 + v_2 f_2 + v_3 f_3$	$v = (v_1, v_2, v_3)$
Planes	$\pi = a F_1 + b F_2 + c F_3 + d F_0$	$ax + by + cz + d = 0$
Lines	$l = (p_2 \wedge p_1)^{\perp}$	$l = (v, (P - O) \times v)$
	$l = \pi_2^{\perp} \wedge \pi_1^{\perp}$	$a_j x + b_j y + c_j z + d_j = 0, \ j = 1, 2$
	$l = l_{32} f_{32} + l_{13} f_{13} + l_{21} f_{21} + l_{01} f_{01} + l_{02} f_{02} + l_{03} f_{03}$	$l = v + u\varepsilon \quad v \cdot u = 0$
	$0 = l_{01} l_{32} + l_{02} l_{13} + l_{03} \ l_{21}$	

Duality (Basis Multivectors)

$$1^\perp = f_{0123} \quad \text{and} \quad f_{0123}^\perp = 1$$

$$f_{ij}^\perp = -f_{0k} \quad \text{and} \quad f_{0k}^\perp = -f_{ij} \quad (ijk \text{ is an even permutation of 321})$$

$$f_j^\perp = F_j \quad \text{and} \quad F_j^\perp = -f_j \quad j = 0,1,2,3$$

$$rr^\perp = f_{0123} \quad \text{(all blades)}$$

$$(r^\perp)^\perp = r \quad \text{(even grades)}$$

$$(r^\perp)^\perp = -r \quad \text{(odd grades)}$$

Meets (Intersection Formulas)

$$p = \left(l \wedge \pi^\perp\right)^\perp \qquad \text{\textit{intersection point p between a line l and a plane }} \pi$$

$$p = \left(\pi_1^\perp \wedge \pi_2^\perp \wedge \pi_3^\perp\right)^\perp \qquad \text{\textit{intersection point p between three planes }} \pi_1, \pi_2, \pi_3$$

$$l = \pi_2^\perp \wedge \pi_1^\perp \qquad \text{\textit{intersection line l between two planes }} \pi_1, \pi_2$$

Joins (Connection Formulas)

$$l = \left(p_2 \wedge p_1\right)^\perp \qquad \text{\textit{bivector representing the line joining two distinct points}}$$

$$\pi = p_1 \wedge p_2 \wedge p_3$$

trivector representing the plane containing three non-collinear points

Incidence Relations

$$p \wedge \pi = 0 \quad \text{\textit{point p lies on plane }} \pi$$

$$p \wedge l^\perp = 0 \quad \text{\textit{point p lies on line l}}$$

$$l \wedge \pi^{\perp} = 0 \quad \text{line } l \text{ lies on plane } \pi$$

$$l_1 \wedge l_2 \neq 0 \quad \text{line } l_1 \text{ skew to line } l_2$$

Distance Formulas

$$\text{dist}(p, \pi) = \left| (\pi \wedge p)^{\perp} \right| \quad \text{from a point } p \text{ to a plane } \pi \text{ with unit normal vector}$$

$$\text{dist}(p, \pi) = \left| (p - q) \cdot v \right|$$

from a point p to a plane π with a point q and unit normal v

$$\text{dist}^2(p, l) = (p - q) \cdot (p - q) - \frac{\left((p - q) \cdot v \right)^2}{v \cdot v}$$

from a point p to a line l containing a point q and parallel to the direction vector v

$$\text{dist}(p, l) = \frac{\left| (p - q) \times v \right|}{|v|}$$

from a point p to a line l containing a point q and parallel to the direction vector v

5.3 Rotors and Versors

Translation *by the distance a in the direction of the unit vector $t = t_1 f_1 + t_2 f_2 + t_3 f_3$*

$$T = 1 - \frac{1}{2} at = e^{-\frac{1}{2} at} \quad \text{translation rotor}$$

$$t = f_0 \alpha t = \left(t_1 f_{01} + t_2 f_{02} + t_3 f_{03} \right) \alpha$$

$$x \mapsto T x T^{-1} \quad \text{on points, vectors, and planes}$$

$$l \mapsto \left(T l^{\perp} T^{-1} \right)^{\perp} \quad \text{on lines}$$

Rotation *by the angle θ about the line v through the origin*

$$R = \cos(\theta/2) + v\,\sin(\theta/2) = e^{\frac{1}{2}v\theta} \quad \text{rotation rotor}$$

$$v = v_1 f_{32} + v_2 f_{13} + v_3 f_{21} \quad \text{unit quaternion}$$

$$x \mapsto RxR^{-1} \quad \text{on points, vectors, and planes}$$

$$l \mapsto \left(Rl^{\perp}R^{-1}\right)^{\perp} \quad \text{on lines}$$

Reflection *in a plane π with unit normal vector*

$$x \mapsto -(\pi x \pi)\alpha \quad \text{on points, vectors, and planes}$$

$$l \mapsto -\left(\pi l^{\perp}\pi\right)^{\perp}\alpha \quad \text{on lines}$$

5.4 Perspective and Pseudo-Perspective

Perspective *from the eye E located at the origin f_0 to the plane π at a distance d from the eye perpendicular to the unit vector **t**.*

$$T = 1 + \frac{1}{2d}t = \text{translation rotor with } t = f_0\alpha t$$

$$P \mapsto \left(T(P-E)^{\perp}T^{-1}\right)^{\perp}$$

Pseudo-Perspective *with the eye located at $E = f_0 - dt$ at a distance d from the origin in the direction −**t***

$$T = 1 + \frac{1}{2d}t = \text{translation rotor with } t = f_0\alpha t$$

$$P \mapsto \left(TP^{\perp}T^{-1}\right)^{\perp}$$

6 Comparisons between the Point Model and the Plane Model of Clifford Algebra

We close by comparing and contrasting some properties of the point model and the plane model of Clifford algebra, and by highlighting as well some of the advantages and disadvantages of each approach.

Algebra

- Plane Model: Natural Algebra, $g_0^2 = 0$
- Point Model: Exotic Algebra, $f_0^2 = 1/\alpha \approx \infty$ {α is an infinitesimal}

Geometry

	Point Model	Plane Model
Points	$p = f_0 + p_1 f_1 + p_2 f_2 + p_3 f_3$	$p = G_0 + p_1 G_1 + p_2 G_2 + p_3 G_3$
Vectors	$v = v_1 f_1 + v_2 f_2 + v_3 f_3$	$v = v_1 G_1 + v_2 G_2 + v_3 G_3$
Planes	$\pi = aF_1 + bF_2 + cF_3 + dF_0$	$\pi = ag_1 + bg_2 + cg_3 + dg_0$
Lines	$l = (p_2 \wedge p_1)^{\perp}$	$l = (p_2^{\perp} \wedge p_1^{\perp})^{\perp}$
	$l = \pi_2^{\perp} \wedge \pi_1^{\perp}$	$l = \pi_2 \wedge \pi_1$

Observations on Geometry

- Plane Model: Exotic geometry—basis vectors g_0, g_1, g_2, g_3 represent planes
- Point Model: Natural geometry—basis vectors f_0, f_1, f_2, f_3 represent points and vectors
- Lines in the point model have natural Plucker coordinates; lines in the plane model have natural dual Plucker coordinates
- Plane Model $\rightarrow$ Point model
  - for points, vectors, and planes: replace g's by f's, and exchange lower case and upper case;
  - for lines: replace each factor by its dual.

Derivations

- Plane Model: Formula for a line as the join of two points is difficult to derive

- Point Model: Formula for a line as the meet of two planes is straightforward to derive

Representations

- The representations of geometry in the two models are almost equivalent, except that the representation for lines as the meet of two planes is much simpler in the plane model.

Incidence Relations

	Point Model	**Plane Model**
point p lies on plane π	$p \wedge \pi = 0$	$\pi \wedge p = 0$
point p lies on line l	$p \wedge l^{\perp} = 0$	$l^{\perp} \wedge p^{\perp} = 0$
line l lies on plane π	$l \wedge \pi^{\perp} = 0$	$\pi \wedge l = 0$
line l_1 skew to line l_2	$l_1 \wedge l_2 \neq 0$	$l_1 \wedge l_2 \neq 0$

Observations on Incidence Relations

- Point on plane and line skew to line are identical in both models.
- Point on line is simpler in the point model.
- Line on plane is simpler in the plane model.

Meets and Joins

	Point Model	**Plane Model**
Line Plane	$\left(l \wedge \pi^{\perp} \right)^{\perp}$	$\pi \wedge l$
Three Planes	$\left(\pi_1^{\perp} \wedge \pi_2^{\perp} \wedge \pi_3^{\perp} \right)^{\perp}$	$\pi_1 \wedge \pi_2 \wedge \pi_3$
Two Planes	$\pi_2^{\perp} \wedge \pi_1^{\perp}$	$\pi_2 \wedge \pi_1$
Two Points	$\left(p_2 \wedge p_1 \right)^{\perp}$	$\left(p_2^{\perp} \wedge p_1^{\perp} \right)^{\perp}$
Three Points	$p_1 \wedge p_2 \wedge p_3$	$\left(p_1^{\perp} \wedge p_2^{\perp} \wedge p_3^{\perp} \right)^{\perp}$

Observations on Meets and Joins

- Meets are simpler in the plane model.
- Joins are simpler in the point model.

Rotors and Versors

	Point Model	Plane Model
Translation Rotor	$T = 1 - \frac{1}{2}at$	$T = 1 - \frac{1}{2}at$
	$t = (t_1 f_{01} + t_2 f_{02} + t_3 f_{03})\alpha$	$t = t_1 g_{01} + t_2 g_{02} + t_3 g_{03}$
Rotation Rotor	$R = \cos(\theta/2) + v \sin(\theta/2)$	$R = \cos(\theta/2) + v \sin(\theta/2)$
	$v = v_1 f_{32} + v_2 f_{13} + v_3 f_{21}$	$v = v_1 g_{32} + v_2 g_{13} + v_3 g_{21}$
Reflection Versor	$\pi = aF_1 + bF_2 + cF_3 + dF_0$	$\pi = ag_1 + bg_2 + cg_3 + dg_0$
	$x \mapsto -(\pi x \pi)\alpha$	$x \mapsto \pi x \pi$

Observations on Rotors and Versors

- The rotors in the two models are almost identical: just exchange *g*'s and *f*'s. But in the translation rotor, the point model introduces a factor of α.
- The versors for reflection are more complicated in the point model than in the plane model. In the point model a minus sign and a factor of α are introduced, and lines *l* are treated as a special case: $l \mapsto -\left(\pi l^{\perp} \pi\right)^{\perp} \alpha$.
- The rotors on lines are much more complicated in the point model than in the plane model.
  - In the point model for a line $l = (p_2 \wedge p_1)^{\perp}$ and a rotor S: $l \mapsto \left(Sl^{\perp} S^{-1}\right)^{\perp}$
  - In the plane model for a line $l = \pi_2 \wedge \pi_1$ and a rotor $S: l \mapsto SlS^{-1}$.

Advantages

- Plane Model
  - i. Natural Algebra: No need to carry along infinitesimals during computations
  - ii. Simpler formulas for transformations on lines
  - iii. Simpler formula for reflection
  - iv. Simpler formulas for meets
  - v. Line on plane simpler in the plane model

- Point Model
    i.   Natural Geometry: Grade 1 basis elements represent points and vectors
    ii. All blades have inverses
    iii. Simpler derivation in the point model of a line as the meet of two planes compared to the derivation in the plane model as a line as the join of two points
    iv. Simpler formulas for joins
    v.   Point on line simpler in the point model

Bibliography

1. Clifford, W.K. 1873. Preliminary sketch of biquaternions. *Proceedings of the London Mathematical Society*, Vol. 4, pp. 381–395.
2. Conway, J. and Smith, D. 2003. *On Quaternions and Octonions: Their Geometry, Arithmetic, and Symmetry*, A.K. Peters, Natick, MA.
3. Daniilidis K. and Bayro-Corrochano E. 1996. The dual quaternion approach to hand-eye calibration. In *Proceedings of ICPR'96*, Vienna, Austria, Vol. 1, pp. 318–322.
4. Dooley, J.R. and McCarthy, J.M. 1991. Spatial rigid body dynamics using dual quaternion components. In *Proceedings. 1991 IEEE International Conference on Robotics and Automation*, Sacramento, CA, USA, Vol. 1, pp. 90–95.
5. Dorst, L. and De Keninck, S. 2022. *A Guided Tour to the Plane Based Geometric Algebra PGA*, Version 2.0, University of Amsterdam, https://bivector.net/PGA4CS.html.
6. Dorst, L., Fontijne, D., and Mann, S. 2007. *Computer Algebra for Computer Science: An Object-Oriented Approach to Geometry*, The Morgan Kaufmann Series in Computer Graphics, Elsevier, San Francisco.
7. Du, J., Goldman, R., and Mann, S, 2016. Modeling Affine and Projective Transformations in 3-Dimensions by Linear Transformations in 4-Dimensions, University of Waterloo Technical Report CS-2018-05.
8. Goldman, R. 2009. *An Integrated Introduction to Computer Graphics and Geometric Modeling*, CRC Press, Taylor and Francis, Boca Raton.
9. Goldman, R. 2010. *Rethinking Quaternions: Theory and Computation*, Synthesis Lectures on Graphics and Animation, Barsky B. (ed), Morgan & Claypool Publishers, California.
10. Grafton, A. and Lasenby, J. 2020. Surface fitting using dual quaternion control points with applications in human respiratory modelling. In *Computer Graphics International Conference, CGI 2020: Advances in Computer Graphics*, pp. 421–433.
11. Gunn, C. 2011. On the homogeneous model of Euclidean geometry. In Dorst, L., and Lasenby, J. (eds). *A Guide to Geometric Algebra in Practice*, pp. 297–327, Springer, London.
12. Gunn, C. and de Keninck, S, 2019. Siggraph 2019 Tutorial on Geometric Algebra. https://www.youtube.com/watch?v=tX4H_ctggYo
13. Gunn, C. and de Keninck, S. 2019. Geometric Algebra for CGI, Vision, and Engineering. https://bivector.net.
14. Hamilton, W. 1866. *Elements of Quaternions*, Cambridge University Press, Cambridge, UK.
15. Hearn, D., Baker, M.P., and Carithers, W. 2011. *Computer Graphics with OpenGL* (4th edition), Pearson, New York.
16. Hunt, M., Mullineux, G., Cripps, R., and Cross, B. 2016. Quaternions, dual quaternions and geometric algebra. Preprint.

17. Hurwitz, A. 1898. Uber die Composition der Quadratischen Formen von Beliebig Vielen Variablen, Nachrichten von der Gesellschaft der Wissenschaften zu Gottingen, Mathematisch-Physikalische, Klasse, Bd. II, pp. 309–316.
18. Juttler, B. 1994. Visualization of moving objects using dual quaternion curves. *Computers and Graphics*, Vol. 18, pp. 315–326.
19. Kavan, L., Collins, S., O'Sullivan, C., and Zara, J. 2006. Dual quaternions for rigid transformation blending. In *Proceedings Siggraph 2006*.
20. Kenwright, B. 2012. A beginners guide to dual quaternions: What they are, how they work, and how to use them for 3D character hierarchies. In *Proceedings WSCG 2012*.
21. Mullineux, G. 2004. Modeling spatial displacements using Clifford algebra. *Journal of Mechanical Design*, Vol. 126, pp. 420–424.
22. Mullineux, G. and Simpson, L. C. 2011. Rigid-body transforms using symbolic infinitesimals. In Dorst, L., and Lasenby, J. (eds). *Guide to Geometric Algebra in Practice*, pp. 353–369, Springer, London.
23. Ozgu, E. and Mezouar, Y. 2016. Kinematic modeling and control of a robot arm using unit dual quaternions. *Robotics and Autonomous Systems*, Vol. 77, pp. 66–73.
24. Price, W., Ton, C., MacKunis, W., and Drakunov, S. 2013. Self reconfigurable control for dual-quaternion/dual-vector systems. In *Proc. 2013 European Control Conference (ECC)*, Zurich, pp. 860–865.
25. Selig, J. 2000. Clifford algebra of points, lines, and planes. *Robotica*, Vol. 18, pp. 545–556.

Index

Note: **Bold** page numbers refer to tables; *italic* page numbers refer to figures.

Printed in the United States
by Baker & Taylor Publisher Services